The Royal Navy's TYPE 45 DESTROYER

C000044328

Acquisition Options and Implications

John Birkler, John F. Schank, Mark V. Arena, Giles Smith, Gordon Lee

Prepared for the United Kingdom's Ministry of Defence
Approved for public release; distribution unlimited

RAND Europe

The research described in this report was prepared for the United Kingdom's Ministry of Defence.

ISBN: 0-8330-3203-8

RAND is a nonprofit institution that helps improve policy and decisionmaking through research and analysis. RAND® is a registered trademark. RAND's publications do not necessarily reflect the opinions or policies of its research sponsors.

The cover art and Figures 1.1, 1.2, 4.4, and 5.1 are from the United Kingdom's Ministry of Defence and are reprinted by permission. The photos in Appendix C are by Tom Lamb. Figure C.4 is from Carlos Merino (IZAR Ferrol Shipyard), "F-100 and F-319: New Frigates for Europe", paper presented at RINA Conference WARSHIP 2001—Future Surface Warships, London, 2001.

Cover design by Maritta Tapanainen

Published 2002 by RAND
1700 Main Street, P.O. Box 2138, Santa Monica, CA 90407-2138
1200 South Hayes Street, Arlington, VA 22202-5050
201 North Craig Street, Suite 202, Pittsburgh, PA 15213-1516
RAND URL: http://www.rand.org/
To order RAND documents or to obtain additional information, contact Distribution Services: Telephone: (310) 451-7002; Fax: (310) 451-6915; Email: order@rand.org

In April 2001, the United Kingdom's Ministry of Defence (MOD) commissioned RAND to investigate procurement strategies that the MOD could pursue as it acquires warships over the next 15 to 20 years. The research was intended both to inform MOD decisions about its next-generation destroyer, the Type 45, and to help the Ministry pursue its long-term warship acquisition programme. In commissioning the research, the MOD asked RAND to evaluate near-term and long-term strategies that would yield the highest value for money, encourage innovation, allow for the efficient use of production capacity, and sustain the United Kingdom's (UK's) core warship industrial base.

This book summarises RAND's analysis of acquisition options open to the MOD during production of the Type 45 Destroyer. This destroyer, to be delivered from 2007, will become a main component in the Royal Navy's surface fleet, taking on roles as diverse as protecting the fleet in littoral settings, participating in hostile engagements on the open ocean, and conducting diplomatic and crisis-intervention missions. The Type 45 also will constitute a large proportion of new-ship acquisitions that the MOD has planned for the next two decades. As such, the acquisition and production techniques that the MOD adopts for the Type 45 will be a bellwether for later acquisitions.

RAND quantitatively and qualitatively evaluated the effect on the United Kingdom's shipbuilding labour force and infrastructure of various options to acquire and build the Type 45. For their evaluations, RAND staff relied on data relating to the future demand for

Royal Navy ships, to commercial work and Foreign Military Sales, and to the existing capacities of the United Kingdom's shipbuilding industrial base. The analysis results were a key input to the UK's decision on the Type 45 programme, as announced to the House of Commons on 10 July 2001, by the Secretary of State for Defence, Mr. Geoffrey Hoon.

This book should be of special interest not only to the Defence Procurement Agency and to other parts of the Ministry of Defence, but also to service and defence agency managers and policymakers involved in weapon system acquisitions on both sides of the Atlantic Ocean. It should also be of interest to shipbuilding industry executives in the United Kingdom. This research was undertaken for the MOD's Chief of Defence Procurement jointly by RAND Europe and the International Security and Defense Policy Center of RAND's National Security Research Division (NSRD), which conducts research for the U.S. Department of Defense, allied foreign governments, the intelligence community, and foundations.

CONTENTS

TABLES

SUMMARY

The United Kingdom has several ship procurement programmes under way or planned for the future, including those for new attack submarines, aircraft carriers, surface combatants, and auxiliary ships. Among these programmes is the new-generation destroyer, known as the Type 45. The United Kingdom intends to acquire up to a dozen of these warships for the Royal Navy beginning in 2007. These ships will be some of the largest surface combatants built for the Royal Navy since World War II and will constitute a sizeable portion of the new surface warship acquisitions that the United Kingdom plans to make over the next two decades. As such, the acquisition and production techniques that the UK Ministry of Defence (MOD) adopts for the Type 45 will be a bellwether for later acquisitions.

In recent years, the MOD has become interested in evaluating a variety of different strategies that it might employ to acquire warships. It also has become increasingly aware of excess industrial capacity in the UK's warship-building sector. As a result, the MOD in 2001 asked RAND to help it (1) analyse the costs and benefits of alternative acquisition paths and (2) evaluate near- and long-term strategies that would yield the highest value, encourage innovation, use production capacity efficiently, and sustain the UK's core warship industrial base, given other current and future MOD ship programmes, with particular reference to the Type 45.

Done quickly at the request of the Chief of Defence Procurement (CDP), this analysis entailed a quantitative comparison of the advantages and disadvantages of having one or two shipbuilding companies produce the Type 45 over the next 15 years. Options examined include employing either whole-ship or modular, so-called block, production techniques in a single shipyard or multiple shipyards. The analysis aimed to help MOD policymakers in two ways: first, to gain an understanding of the costs and benefits of different Type 45 acquisition and production strategies; and second, to gauge the effect of those strategies both on the United Kingdom's shipbuilding industrial base and on the costs of other current and future MOD ship programmes.

HOW CAN MOD MOST EFFECTIVELY MAINTAIN A COMPETITIVE PRODUCTION ENVIRONMENT?

The MOD's stated policy is to make maximum use of competition whenever possible, based on the theory that competition promotes innovation, reduces costs, and sustains core industrial capabilities. Therefore, MOD generally views acquisition and production strategies that allow for robust, sustained competition in a more favourable light than alternatives that rely on less competitive arrangements. However, in some situations (detailed in Birkler et al., 2001), the use of multiple, competitive sources can incur some penalties, especially in the near term. Careful, situation-specific analysis is needed to identify the best course of action.

The Ministry's original conception for competitive acquisition of the Type 45 called for the involvement of two companies, BAE SYSTEMS Marine[1] and Vosper Thornycroft (UK) Limited (VT),[2] in manufacturing the destroyer. Both companies were to share in producing the first three ships and compete to build subsequent batches of the vessel.

---

[1]BAE SYSTEMS Marine is a subsidiary of BAE SYSTEMS, a UK-based company that designs and manufactures civil and military aircraft, surface ships, submarines, space systems, radar, avionics, communications, electronics, guided-weapon systems, and a range of other defence products. Formed by the merger of British Aerospace and Marconi Electronic Systems in November 1999, BAE SYSTEMS employs more than 100,000 people worldwide and has annual sales of some £12 billion.

[2]Vosper Thornycroft is a UK-based company that focuses on design, development, and production of warships, and commercial and military marine electronic controls; and provides management, technical, training, and educational support services to military and civil markets. Together, these activities employ 7,500 people in a network covering the UK, Europe, the United States, and the Middle East.

This strategy required BAE SYSTEMS Marine and VT to cooperate in designing and producing the first three ships. Their cooperation, referred to as "the Alliance", succeeded initially, then foundered when prices and terms acceptable to all parties could not be agreed on.

The strategy was further challenged in December 2000, when BAE SYSTEMS Marine submitted an unsolicited proposal to the MOD to build all 12 Type 45 Destroyers. BAE SYSTEMS Marine argued that producing the destroyers in its own shipyards would enable it to deliver the ships faster and at lower cost than it could through a joint-production arrangement with VT.

The MOD's original joint-production strategy and BAE SYSTEMS Marine's subsequent sole-source proposal came about in a domestic industrial context characterised by excess commercial and naval shipbuilding capacity. Over the past 20 years, shipyards in the United Kingdom have seen demand drop as orders for commercial ships have gone to lower-cost foreign competitors. The result: almost every significant UK shipyard has at one time or another seen its capacity exceed demand.

RAND'S THREE INTERWOVEN QUESTIONS

RAND researchers qualitatively and quantitatively addressed three interwoven questions:

- *One producer or two producers?* Should the MOD have the Type 45 built by one company or by two?

- *Competitive or directed allocation of work if two producers?* Should the MOD compete the 12 ships in the Type 45 class, recognising that competition may result in a single producer, or should it directly allocate work to specific shipbuilders to ensure that both producers stay involved with the programme?

- *Whole-ship or block production?* Should the company or companies producing the Type 45 construct the destroyer in its entirety in one shipyard or assemble it from segments or blocks produced in several shipyards?

Answers to these questions have implications not just for the Type 45 but also for other MOD warship acquisition programmes, for a number of UK shipyards, and for the UK's pool of skilled shipyard work-ers. Depending on how the MOD decides to acquire, produce, main-tain, and repair the Type 45, one or more UK shipbuilders could see their revenues drop, causing the closure of one or more shipyards.

HOW RAND STUDIED THE PROBLEM

RAND researchers developed an analytic model of the UK shipbuild-ing industrial base specifically for this Type 45 study. The model encompassed all current and future programmes at BAE SYSTEMS Marine and VT shipyards and calculated workforce, overhead, and investment costs.

Using as its baseline the option of having BAE SYSTEMS Marine build all 12 Type 45s, the model displayed the relative-cost effects of alternative procurement paths on the Type 45 and other pro-grammes.

To assess the direct-cost consequences of spreading production be-tween two firms, as would occur in a strategy involving continuing competition (rather than concentrating it at one firm), we estimated the labour costs, overhead rates, labour-force transition costs, and learning improvements across several shipyards for a given ship-building strategy. We then separately estimated the likelihood that competitive pressures would reduce costs enough to overcome the cost penalties of distributing work among two producers. *Thus, the basic question is not how much money will be saved but, rather, whether introducing an additional production source is a reasonable strategy to pursue.*

This model required RAND to collect extensive data of the type out-lined in Table S.1.[3] We collected this information from discussions held during site visits with five UK shipbuilders: VT, BAE SYSTEMS Marine, Swan Hunter, Appledore, and Harland and Wolff. Beyond

---

[3]Our study excluded costs of the weapon system and other Government-Furnished Equipment, and the cost of common items and material purchased by the Prime Con-tract Office (PCO) and provided to the shipyards.

Table S.1

Data Used in RAND Shipyard Model

Data	Description
Shipyard capacity	Steel throughput, docks, lifting capacity, outfitting[a] berths, and the like
Workforce profile	Experience levels, age, productivity, costs for hiring and training, termination costs, restrictions on hiring and termination, current employment levels, and the use of contract workers
Production experience	Numbers and types of ships built over the past five years (including commercial work)
Current and future production	Current and anticipated production plans (by ship)
Workload projections	For each activity listed in the current and future production plan, a listing of the labour profile by trade and quarter. Further, these data included design and development workload.
Wage rates	Wage rates for all the labour types
Burden rates	Overheads, General and Administrative, and profit rates as a function of different site workloads
Investment levels	Fixed investments, such as facilities, necessary for a particular programme or that investment required overall

[a]*Outfitting* includes installation of furnishings (e.g., desks and chairs) and equipment (e.g., electrical systems and pipes).

RAND *MR1486-TS.1*

the shipbuilders, RAND researchers had similar discussions with Type 45 and other MOD ship programme managers and their staffs. These offices provided considerable supporting information and data on their ship programmes.

WHAT RAND FOUND OUT

The sole-source option of building all 12 ships in one yard is calculated to be less costly than a direct-allocation strategy based on capital investment required, labor productivity, work process, and overhead savings. A direct-allocation strategy would involve dividing the total production between two shipyards in some predetermined way. We studied three possible distributions—eight, six, or four ships—to BAE SYSTEMS Marine, with the remaining ships going to

VT. These cases resulted in an increase in programme costs of 10 to 13 percent over the baseline of BAE SYSTEMS Marine building all 12 ships (see Figure S.1).

A competitive strategy is similar to a direct-allocation strategy in that the total production would be divided between the two shipbuilders. However, there are two important differences. First, the final distribution of work is unknown; that is, any of the distributions studied is possible. Second, competitive forces would reduce the price of the ships over those for a direct-allocation strategy. The question is, Could competitive forces overcome the 10 to 13 percent increase over the sole-source case? To answer this question, we next estimated the likelihood that competitive forces sustained throughout the production programme could overcome these cost increases. Historical data, collected on 31 ship and missile programmes that involved head-to-head competition on some part of the production phase, suggest an approximately even chance that competition would lower production costs by at least 13 percent. Therefore, we estimate roughly an even chance that competitive production of the Type 45 at two shipyards would yield about the same overall cost as sole-source production at one shipyard. We can give no definite answer on whether competitive or sole-source production would most likely lead to lower costs for the projected 12-ship production programme.

Even with competition, there is the chance that one shipbuilder would end up building most of the ships or all of the last competitive batch. An alternative acquisition strategy to ensure that both shipbuilders would remain in the UK warship-building industry is to direct a certain number of ships, or portions of the ships, to each of the two shipbuilders. Directed buys of whole ships to the shipbuilders result in the 10- to 13-percent cost increases for the competitive options, but without the potential for lower costs due to competition. However, allowing each shipbuilder to build the same sections, or blocks, of all 12 ships not only keeps both companies involved in building warships but also takes maximum advantage of the lower production man-hours due to learning.[4]

_____

[4]We recognise that this strategy might not preserve the shipbuilding skills unique to the blocks produced at the other shipyard.

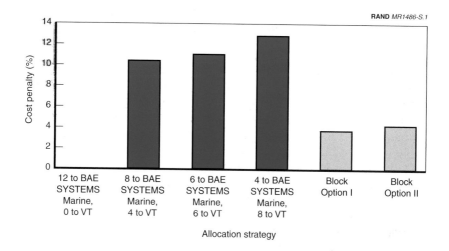

RAND *MR1486-S.1*

Figure S.1—Summary Comparison of Cost Penalties
of Six Procurement Options

We examined two different options involving the distribution of the
various blocks to the different shipyards. These two block options
resulted in cost increases of approximately 4 percent over the sole-
source option of having BAE SYSTEMS Marine build all 12 ships.

RAND'S FINDINGS ALREADY HAVE BEEN PUT TO USE

The analysis described here was presented to senior MOD leadership
during the first half of June 2001 and was a key input to the United
Kingdom's decision on the way forward. On 10 July, the Secretary of
State for Defence, Mr. Geoffrey Hoon, announced to the House of
Commons that BAE SYSTEMS Marine and VT would jointly build
Type 45 destroyers in blocks. Mr. Hoon subsequently announced, on
18 February 2002, that the MOD had made a contractual commitment with the prime contractor for a further three ships (in addition
to the three already on order) and that the prime contractor had, in
turn, committed with BAE SYSTEMS Marine and VT for their work on
the first six Type 45 platforms in the planned class of up to 12.

This solution carries a number of advantages. Spreading the Type 45
work between BAE SYSTEMS Marine and VT helps ensure that both

shipbuilders will remain viable and able to compete on future MOD programmes such as the Future Surface Combatant (FSC). Having a single shipbuilder construct all of the same blocks for the Type 45 class takes maximum advantage of the learning effect and, therefore, reduces costs below those of distributing the construction of complete ships between the two shipbuilders.

Moreover, the revised Type 45 strategy will allow VT to proceed with plans to invest in a new shipbuilding facility within the Portsmouth Royal Naval Base. Since spring 2000, the company has been planning to shift shipbuilding operations to Portsmouth, but had delayed committing to this move pending the MOD's determination of VT's role in the Type 45 programme. Shortly after the MOD's decision, VT announced that it would go ahead with the new facility, to be built on four existing docks in the Portsmouth Royal Naval Base. VT has signed a 125-year lease for part of the base. The lease will come into effect shortly.

The MOD's decision is not risk-free. Building ship blocks at multiple sites has potential disadvantages. Constructing the blocks with enough rigidity and weatherproofing to permit movement and transportation will carry additional costs. Structural tolerances must be managed very closely, since any misalignment of adjacent blocks can lead to substantial rework costs. Also, block-transportation costs might be higher than we assumed. Finally, scheduling of the construction and delivery of the blocks must be closely managed. A block that arrives late at the assembly yard may cause significant delays. We identified such risks but were unable to quantify their effect. On balance, we found no persuasive evidence to suggest that the potential penalties of such risks outweighed the potential benefits of the block-construction strategy.

ACKNOWLEDGEMENTS

This work could not have been undertaken without the steadfast support and encouragement we received from Sir Robert Walmsley, Chief of Defence Procurement & Chief Executive, Defence Procurement Agency (DPA), and members of his staff. Many individuals in the UK Ministry of Defence (MOD) provided their time, knowledge, and information to help us perform the analyses discussed in this book. Their names and contributions would fill several pages.

If we were to single out two persons who participated in and supported this work in extraordinary ways, we would mention Charles Draper, of the MOD, our action officer, and Brigadier Keith Prentice, Integrated Project Team (IPT) leader for the Type 45 Destroyer. Their tireless efforts on our behalf are greatly appreciated, and their friendship is even more valued.

R Adm Nigel Guild, of the DPA, also deserves special mention—and our gratitude. His insights and questions were highly valued and caused us to think more broadly on the issues. Nick Witney, Director General Equipment, graciously described the programming process to us. These interactions added to the richness and quality of the study.

We also want to thank the leadership and staff of MOD, especially Ali Baghaei, Simon Rusling, Karl Monk, Cdre Dave Maclean RN, Terry Baldwin, and Roger Russell, the IPT leaders for MOD's current and future ship and submarine programmes, whom we interviewed during this study. As well, we thank Terry Johns, Group Pricing Manager, Land and Naval Systems; and Arthur Fisher, Cdre Neil Latham RN , and John Hall, Warship Support Agency. They and their staffs gave generously of their time and advanced our understanding of the UK shipbuilding programmes and industrial capabilities.

We are also indebted to Brian Phillipson, then Managing Director, Type 45 Prime Contract Office, BAE SYSTEMS; Nigel Whitehead, then Managing Director, Astute Class Submarine Prime Contract Office, BAE SYSTEMS; J. R. Wilson, Managing Director, Appledore Shipbuilders Ltd.; Andrew Bunney, Managing Director–Shipbuilding, Vosper Thornycroft (UK) Limited; Simon Kirby, then Managing Director, BAE SYSTEMS Marine Division; Jaap Kroese, Chairman, Swan Hunter; and Jim Gregg, Business Development Manager, Harland and Wolff Heavy Industries Ltd. Each gave us the opportunity to discuss a broad range of issues with those directly involved. In addition, all of the firms arranged for us to visit their facilities. The firms and government offices provided all the data we requested in a timely manner.

Finally, for their careful and constructive comments on earlier drafts, we wish to thank Charles Draper and Andy McClelland, MOD, and RAND colleagues Frank Lacroix and Irv Blickstein. We are additionally indebted to Joan Myers for her deft assistance in organising and formatting the many drafts, and to Marian Branch for her editing.

ACRONYMS

AAW	Anti-air warfare
ALSL	Alternative Landing Ship Logistics
AO	Auxiliary Oiler
AOR	Fleet Replenishment Ship
ASW	Anti-submarine warfare
CDP	Chief of Defence Procurement
CNGF	Common New Generation Frigate
CPC	Competing Prime Contractor
CVF	Future Aircraft Carrier
D&M	Demonstration and Manufacture
DD	Destroyer
DDH	Devonshire Dock Hall
DFM	Demonstration and First of Class Manufacture
DPA	Defence Procurement Agency (formerly the Procurement Executive)
FF	Frigate
FOC	First-of-class
FOPV	Future Offshore Patrol Vessel
FPSO	Floating Production, Storage and Offloading vessel

FSC	Future Surface Combatants
FSED/IP	Full-Scale Engineering Development and Initial Production
FSL	Fleet Support Limited
G&A	General and Administrative
GEC	General Electric Corporation
GFE	Government-Furnished Equipment
GRT	Gross registry tonnage
HM&E	Hull, Mechanical and Electrical
IPT	Integrated Project Team
JCB	Joint Capability Board
LBP	Length Between Perpendiculars
LCU	Landing Craft Utility
LCVP	Landing Craft Vehicle Personnel
LNG	Liquid nitrogen gas
LPD	Landing Platform Dock
LPD(R)	Landing Platform Dock (Replacement)
LSL	Landing Ship Logistics
MCMV	Mine Countermeasures Vessel
MIL	Marine Industries Limited
MOD	Ministry of Defence
MOU	Memorandum of Understanding
NAPNOC	No Acceptable Price/No Contract
OH	Overhead
OPV	Offshore Patrol Vessel
PAAMS	Principal Anti-Air Missile System
PCO	Prime Contract Office

PFI	Private Finance Initiative
R&D	Research and Development
RAF	Royal Air Force
RFA	Royal Fleet Auxiliary
RN	Royal Navy
RoRo	Roll-on–Roll-off ship
SDR	Strategic Defence Review
SSA	Shipbuilders and Shipbreakers Association (UK)
SSBN	Nuclear-powered ballistic-missile submarine
SSK	Conventional submarine
SSN	Nuclear-powered attack submarine
UK	United Kingdom
U.S.	United States
VLCC	Very large crude carrier
VSEL	Vickers Shipbuilding and Engineering Limited
VT	Vosper Thornycroft (UK) Limited

Chapter One

INTRODUCTION

Between April and September 2001, RAND researchers analysed options open to the Ministry of Defence (MOD) in the United Kingdom to acquire and produce the Royal Navy's (RN's) next-generation destroyer, the Type 45. Done quickly at the request of the Chief of Defence Procurement (CDP), this analysis entailed a quantitative comparison of the advantages and disadvantages of having either one or two shipbuilding companies produce the Type 45 over the next 15 years. Options examined included employing either whole-ship or modular, so-called block, production techniques, either in a single shipyard or in multiple shipyards. The analysis aimed to help MOD policymakers in two ways: first, to gain an understanding of the costs and benefits of different Type 45 acquisition and production strategies; and, second, to gauge the effect of those strategies both on the United Kingdom's shipbuilding industrial base and on the costs of other current and future MOD ship programmes.

The MOD's policy is to pursue competition in defence procurement, at either the prime-contract or subcontract level, in order to secure value for money. The MOD generally views acquisition and production strategies that allow for robust, sustained competition in a more favourable light than alternatives that rely on less competitive arrangements. This preference is based on the premise that competition promotes innovation, reduces costs, and sustains core industrial capabilities. However, in some situations (detailed in Birkler et al., 2001), the use of multiple, competitive sources can incur some penalties, especially in the near term. Consequently, careful, situation-specific analysis to identify the best course of action is needed.

BACKGROUND AND STUDY OBJECTIVES

The MOD has several future ship procurement programmes, including plans for new attack submarines, aircraft carriers, the Type 45 Destroyer, and auxiliary ships. MOD's plans call for the Royal Navy to take delivery of up to 12 of the destroyers from 2007. The MOD has attempted to make the acquisition of these ships a competitive process. BAE SYSTEMS is the Prime Contract Office (PCO), with overall responsibility for the design and manufacture of the Type 45s.

The original conception of the Type 45 warship acquisition process developed with the PCO called for two companies, BAE SYSTEMS Marine[1] and Vosper Thornycroft (VT),[2] to be involved in the production of the destroyer. These two competitors were to share in the production of the first three ships: Each would build one complete Type 45, and each would build one-half of the third, which would be assembled at BAE SYSTEMS Marine. For the planned next batch of three ships, competition for assembly would be limited to BAE SYSTEMS Marine and VT; the loser would have the option of assembling one of the three ships at the winner's price.

The Alliance Forms

This strategy required BAE SYSTEMS Marine and VT to operate together in the design process, ensuring that the ship was buildable by both shipbuilders. Initially, BAE SYSTEMS Marine and VT formed an "Alliance" for the production of the Type 45. The Alliance did pro-

---

[1]BAE SYSTEMS Marine is a subsidiary of BAE SYSTEMS, a UK-based company that designs and manufactures civil and military aircraft, surface ships, submarines, space systems, radar, avionics, communications, electronics, guided-weapon systems, and a range of other defence products. Formed by the merger of British Aerospace and Marconi Electronic Systems in November 1999, BAE SYSTEMS employs more than 100,000 people worldwide and has annual sales of some £12 billion. One condition of the merger was that a clear firewall be established between BAE SYSTEMS and BAE SYSTEMS Marine, to preclude any bias or favouritism in BAE SYSTEM's award of shipbuilding contracts in a competitive environment.

[2]Vosper Thornycroft is a UK-based company that focuses on design, development, and production of warships and paramilitary craft ranging in size from destroyers and frigates to inshore patrol craft, and commercial and military marine electronic controls; and provides management, technical, training, and educational support services to military and civil markets. Together, these activities employ 7,500 people in a network covering the UK, Europe, the United States, and the Middle East.

duce a price to the prime contractor.[3]   However, the Alliance foundered because, among other things, the shipbuilders could not agree on prices and terms acceptable to the prime contractor. The prime contractor was also concerned about taking on more of the risk and responsibility for the design and build than was originally envisioned.

The Alliance Dissolves

Late in 2000, BAE SYSTEMS Marine informed the MOD that it had concluded from a recently completed ten-year business plan that its three shipyards (Barrow-in-Furness, Scotstoun, and Govan) could not remain viable without increased throughput and restructuring.[4] In December 2000, BAE SYSTEMS Marine sent the Type 45 PCO an unsolicited proposal to build all 12 Type 45 Destroyers as currently designed, with participation by VT. However, the design would be optimised for build at BAE SYSTEMS Marine yards. BAE SYSTEMS Marine argued that by producing the destroyers in its own shipyards, it could deliver the ships faster and at lower cost than under the Alliance strategy.[5]

The Domestic Industrial Context

Both the Alliance strategy and the unsolicited proposal came about in a domestic industrial context characterised by excess commercial and naval shipbuilding capacity. Over the past 20 years, shipyards in the United Kingdom have seen demand drop as commercial orders have gone to lower-cost foreign competitors and fewer naval ships have been ordered. The result: almost every significant UK shipyard has at one time or another seen its capacity exceed demand.

---

[3]In the United Kingdom, the development of particular weapon systems generally falls to a PCO, which is responsible for the entire programme from design through production and, more recently, sometimes a portion of the in-fleet support. This acquisition strategy reduces the risks borne by the MOD. For the Type 45, the PCO is a part of BAE SYSTEMS, the parent company of BAE SYSTEMS Marine.

[4]BAE SYSTEMS Marine now manages Govan and Scotstoun as one "Clyde" facility.

[5]In addition to the 12 Type 45 ships, the unsolicited proposal included other MOD contracts at other BAE SYSTEMS Marine sites.

Such excess capacity, along with a string of corporate mergers, has changed the face of the UK shipbuilding industry, particularly with respect to naval shipbuilding. The building of Royal Navy vessels, which in years past involved numerous small and large shipyards, is today concentrated at two producers, BAE SYSTEMS Marine and VT:

- BAE SYSTEMS Marine has some 90 percent of the naval ship-building capacity and owns three shipyards—Govan and Scots-toun, in Clyde, Scotland; and Barrow-in-Furness, in north-western England. Barrow-in-Furness is the sole UK facility for building nuclear-powered submarines, although substantial new refueling facilities are being completed at Devonport, in southern England.

- Vosper Thornycroft has supplied 270 warships to 34 countries over the past 30 years (including mainly smaller warships for the Royal Navy).

Over the past ten years, several other commercially oriented ship-yards—including Swan Hunter, Harland and Wolff, Appledore, and Cammell Laird—have competed to produce military or auxiliary vessels. The repair shipyards at Devonport, Rosyth, and Portsmouth have not undertaken new construction for many years, but they play the major role in the approximately £300-million annual ship refit and repair programme.

RAND Is Asked to Step In

Faced with different acquisition approaches for the Type 45 (and aware that the approaches had longer-term implications), and with excess industrial capacity, the MOD asked RAND to help it analyse the costs and benefits of alternative acquisition paths. In providing such assistance, RAND qualitatively and quantitatively addressed three interwoven questions:

- *One producer or two producers?* Should the MOD acquire Type 45s built by one company or two?

- *Competitive or directed allocation of work if two producers?* Should the MOD compete some or all of the 12 ships in the Type 45 class, recognising that competition may result in a single producer, or should it directly allocate work to specific shipbuilders to ensure that those producers stay involved with the programme?

- *Whole-ship or block production?* Should the company or companies producing the Type 45 construct the destroyer in its entirety in one shipyard or assemble it from segments, or blocks, produced in several shipyards?

The answers to these questions have implications not just for the Type 45 but also for other MOD warship acquisition programmes, for a number of UK shipyards, and for the UK's pool of skilled shipyard workers. Depending on how the MOD decides to acquire, produce, maintain, and repair the Type 45, one or more UK shipbuilders could see their revenues drop, thereby threatening the future of one or more shipyards.

TYPE 45 DESTROYER

The Type 45 will be one of the largest UK ship programmes in recent history (see Figure 1.1). First-of-class, HMS *Daring*, is due to enter service in 2007. The total cost for the first six ships, including past development costs and UK's share of the separately procured Principal Anti-Air Missile System (PAAMS), is some £4.3 billion. The MOD plans to buy up to 12 of the destroyers.

The Type 45 will replace the Royal Navy's current anti-aircraft destroyer, the Type 42, which is ageing and costly to operate. It will be longer and wider, and will have a larger displacement than the Type 42. As shown in Figure 1.2, it also will have almost twice the displacement of the most recent surface combatant to enter the UK fleet, the Type 23. This larger displacement limits where the Type 45 can be assembled and launched.

This new destroyer is designed for multiple roles.

On the one hand, it will act as the backbone of the Royal Navy's anti-air warfare (AAW) force-protection capability through 2040. In this capacity, its primary mission will be to help UK forces control airspace during joint operations in littoral theatres. Its top speed of 29 knots, its sophisticated array of PAAMS surface-to-air missiles,

RAND *MR1486-1.1*

SOURCE: Graphic posted on Royal Navy's web site at
www.royal-navy.mod.uk/static/content/data/gallery/full/989575547f.jpg.

Figure 1.1—Artist's Rendering of the Type 45 Destroyer

and its SAMPSON radar give it much more robust AAW capabilities
than its predecessors, and its ability to operate a Merlin helicopter
gives it improved anti–submarine warfare (ASW) capabilities.

On the other hand, the Royal Navy also intends to task the Type 45
with protecting maritime trade routes and shipping, and to conduct
open-ocean warfare. The Type 45 must be able to carry out these
missions in the presence of submarine and surface threats and be in-
teroperable with NATO and Allied units. Moreover, the MOD intends
to deploy the Type 45 worldwide in support of broad British interests,
ranging from defence diplomacy to disaster relief to crisis inter-
vention.

RAND *MR1486-1.2*

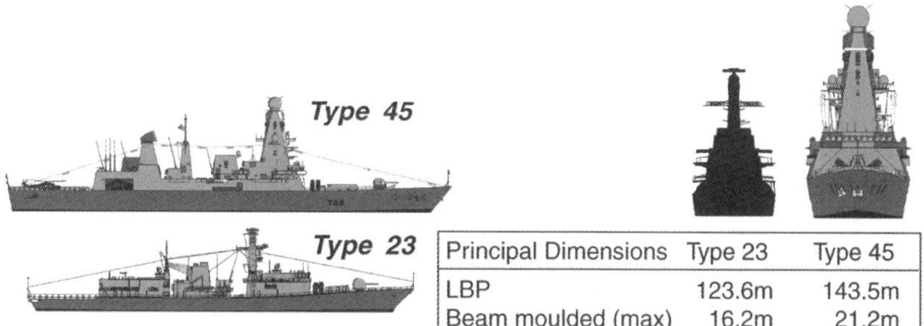

Type 45

Type 23

Principal Dimensions	Type 23	Type 45
LBP	123.6m	143.5m
Beam moulded (max)	16.2m	21.2m
Depth (1 Dk)	8.9m	12.6m
Draught	4.5m	5.65m
Air draught	28.0m	39.0m
Displacement	4300 te	7800 te

Figure 1.2—Relative Size of the Type 45 Destroyer and the Type 23
Destroyer

A £1.2-billion contract for the Demonstration and First of Class Manufacture (DFM) of the Type 45 was let in December 2000 to BAE SYSTEMS PCO. The contract covered the design, development, and delivery of the first three ships, together with elements of support for the first three ships of the class. The contract assumes a work-sharing arrangement between BAE SYSTEMS Marine and Vosper Thornycroft to ensure the credibility of future competition in subsequent batches.

SOURCES OF EVIDENCE WE DREW UPON TO MEET THE STUDY OBJECTIVES

Our analysis was based on four main sources of information, as follows:

- First, the prime contractor teams for Type 45 and Astute pro-grammes[6] provided their own (proprietary) estimates of devel-

---

[6]Data on the Astute programme were needed to understand the total demand for labour and facilities at Barrow-in-Furness, the BAE SYSTEMS Marine shipyard that constructs the Astute submarines.

opment and production costs, at a level of detail that enabled us to estimate costs under different production scenarios.

- Second, several MOD authorities provided their own cost estimates and/or overall programmatic information on past and projected programme schedules, production quantities for each ship class, etc. These sets of quantitative data were augmented by extensive discussions with the PCOs for the Type 45 and for other warships, and with MOD staff and shipyards regarding the feasibility and desirability of various competition strategies.

- Third, each of the shipyards provided extensive historical programme and cost data for all the classes of ships they had produced. They also shared with us their estimates of the costs for future ship programmes.

- Fourth, we conducted a literature review on the results of prior efforts to introduce competition to a weapons-production programme. Because of the short time available for this study (about six weeks from go-ahead to first Type 45 briefing), our analysis of the historical record on the effects of competition on production cost drew on previous studies of production-cost changes due to competition.[7] A more extensive and thorough review of UK acquisition programmes would have been desirable; we have no evidence that a useful body of additional data could be developed.

LIMITATIONS ON THE SCOPE OF THE STUDY

Competition is widely expected to stimulate a variety of actions by the producers in an attempt to make their product more attractive to the buyer. This study was almost entirely focused on a single consequence of competition: changes in production costs. Several other possible consequences of competition are briefly reviewed and factored into the overall conclusions, but only the costs of production are examined quantitatively.

---

[7]The data on historical programmes are extensively described in Table 5.8 and the associated text in John Birkler et al., *Assessing Competitive Strategies for the Joint Strike Fighter*, Santa Monica, Calif.: RAND, MR-1362-OSD/JSF, 2001, pp. 53–56.

Unfortunately, RAND has no historical or analytic methods for directly estimating reductions in production cost due to the introduction of competition. Instead, we estimated the incremental costs of introducing competition. Then, drawing on historical evidence, we attempted to estimate the likelihood that competition would drive down the costs enough to permit recovering those incremental costs, thus allowing the government to at least break even on costs.

ORGANISATION

The book is organised in six chapters. Following this Introduction, Chapter Two sets the overall context in which the Type 45 programme must be executed. It describes the current status of the UK shipbuilding industrial base, and the current and future programmes of the MOD. It then discusses how those programmes are currently matched, or could be matched, to the shipbuilders that make up the industrial base. Chapter Three details the choices facing the MOD for the Type 45 programme and how we quantitatively and qualitatively analysed the various options. Chapter Four presents the results of our analysis of different strategies for distributing the workload for the Type 45 programme between the two shipbuilders. In it we consider strategies that involve the construction of both complete ships and major portions, or blocks, at different shipyards. Chapter Five discusses the Type 45 decision that was made after our analysis concluded. Chapter Six describes the implications of the Type 45 decision for future MOD shipbuilding programmes and the shipbuilding industrial base.

Several appendices provide more detailed descriptions of the material contained in the body of the book: Appendix A describes a sensitivity analysis of the assumptions made in the study. Appendix B reports the long-term consequences of a lack of competition. Appendix C discusses the implications for the Type 45 programme of building ships in blocks at multiple sites.

Chapter Two

THE SHIPBUILDING INDUSTRIAL BASE AND THE MOD SHIPBUILDING PROGRAMME

Several factors constrain the options available to the MOD for producing the Type 45 Destroyer and other military ships. They include the present status and condition of the shipbuilding resources in the UK. Also, the Type 45 programme is but one element of a broader, long-range programme to upgrade the combatant and support forces of the Royal Navy. Any policies and decisions regarding the Type 45 programme must be formulated in recognition of that broader programme.

In this chapter, we first describe the past (first 70 years of the twentieth century) and present UK shipbuilding industry structure and capabilities, outline key economic and political forces affecting the industry, and provide a foundation for a subsequent discussion on how different procurement strategies for the Type 45 programme could affect future industry capabilities. We then present a brief overview of the present fleet, summarise the near- and far-term MOD planning assumptions for Royal Navy ships, and outline the ship-production schedule planned to support those requirements over the next decade. This overview includes a more detailed description of the Type 45 system and overall programme than that given earlier in this book. We conclude the chapter by showing the current and potential distributions of the MOD programmes to the various shipyards.

SHIPBUILDING IN THE UNITED KINGDOM: 1900 TO 1970[1]

Whereas the United Kingdom today constructs less than 1 percent of the world's ships, the country has had a long connection with ship-building. Throughout the nineteenth century and the first two decades of the twentieth century, UK shipyards produced from one-half to three-quarters of the world's shipbuilding output, including a wide range of commercial and military ships.

World War I closed most export markets to British shipbuilders, which never recovered that lost market after the war. Exports ac-counting for nearly 25 percent of UK shipbuilding production before the war constituted only 8 percent of the industry's output after the war. By 1918, the decline in British shipyards had accelerated, as domestic ship owners started to go abroad for their ships. Briefly af-ter 1918, the UK accounted for 50 percent of total world shipbuilding; then, in the 1920s, the proportion and, more important, the absolute total, dropped. Gross tonnage produced fell from over 2 million in 1920 to 0.5 million in the mid-1930s. As a result, many UK shipyards closed during the 1930s.

Just before the start of World War II, British shipbuilding's share of total world output had dropped to 35 percent. After the war, Britain's share fell further, to 15 percent, even though world output more than doubled. Apart from a very brief period in the late 1950s, the UK's shipbuilding industry did not really recover. The high percentage of UK building in 1950 reflected the time it took for European and, in particular, Japanese shipbuilding to recover from World War II.

---

[1]Much of the material in this section and the next was prepared by Tom Lamb, P.E., EUR ING, Technical Associate, Innovative Marine Product Development, LLC, Ann Arbor, Michigan. It was drawn from Anthony Burton, *The Rise and Fall of British Shipbuilding*, London: Constable and Company Limited, 1984; and Robert J. Winklareth, *Naval Shipbuilders of the World: From the Age of Sail to the Present Day*, London: Chatham Publishing, 2000.

SHIPBUILDING IN THE UNITED KINGDOM: 1970 TO 2000

The first oil crisis, nationalisation, privatisation, and consolidation characterised the UK shipbuilding industry during the last three decades of the twentieth century.

The first Arab oil embargo in 1973 caused a worldwide crisis in shipbuilding, driving up fuel prices and reducing demand for ocean transport of all kinds. The result of the lower demands was devastating to the UK shipbuilding industry, forcing several UK shipbuilders close to bankruptcy.

This crisis coincided in the United Kingdom with a change of government. The incoming Labour government was committed to nationalising many industries, a commitment reflecting a view that government had a major role to play in supporting industry. The level of security enjoyed by workers had reached its highest point in the late 1960s. The government decided that the best way to maintain an effective shipbuilding capability in Britain was to nationalise it.

But by the time the industry was nationalised and the structures were in place for the new British Shipbuilders Corporation, the government was again poised to change. In 1979, the Conservatives again took power. The newly nationalised industry was now faced with a government that had the objective of returning shipbuilding to the private sector as soon as possible.

Although the naval shipbuilding sector was relatively strong from 1980 to 1990, owing to Royal Navy requirements and some export work, the UK commercial market had almost dried up. As a result, the government sought to sell or close a number of shipyards. Closures included Cammell Laird, which had been sold by British Shipbuilders to Vickers Shipbuilding and Engineering Limited (VSEL; now BAE SYSTEMS Marine), and a number of smaller shipyards. Foreign management took over other shipyards. Kvaerner of Norway, as part of the development of a large international shipbuilding group, took on the Govan shipyard. Harland and Wolff was taken on by a mix of management and Norway's Olsen group.

Closures continued during the 1990s and included Cochrane in 1993, Richard Dunston in 1994, and Yorkshire Dry Dock in 1998. Swan Hunter, a much larger yard, closed in 1993 after it failed to secure a major naval contract. However, residual demand prompted some of the closed shipyards to reopen later in the decade. Swan Hunter was reopened in 1995 by THC Fabricators, a Dutch firm, and now has secured commercial contracts, including the conversion of a Floating Production, Storage and Offloading (FPSO) vessel, and a naval order for two Alternative Landing Ships Logistics (ALSL). Yorkshire Dry Dock reopened as part of the George Prior Group.

The industry had greater success in the repair and conversion market. Cammell Laird reopened and operated with this market focus from 1993 to 2001, when it went into receivership. At the time of this writing, Cammell Laird at Tyneside is owned by A&P; its yards on Wearside and Teeside have closed.

The change in ownership of the three BAE SYSTEMS Marine shipyards is a prime example of the turmoil in the UK shipbuilding environment during the past two decades. In 1985, the Yarrow shipyard on the Clyde (now Scotstoun) was purchased by GEC-Marconi, the defence arm of the British General Electric Corporation (GEC). In 1986, Vickers Shipbuilding and Engineering Limited, owner of the Barrow-in-Furness shipyard and previously a part of the state-owned British Shipbuilders, was the object of an employee buyout. In 1988, British Shipbuilders sold the Govan shipyard, formerly known as Fairfields, to Kvaerner, a Norway-based conglomerate with several shipyards located throughout the world. In 1994, GEC-Marconi purchased VSEL and incorporated the Barrow-in-Furness and Yarrow shipyards as Marconi-Marine. In 1999, Marconi-Marine bought the Govan yard when Kvaerner announced its exit from the shipbuilding industry. Finally, in September 1999, British Aerospace bought GEC's Marconi Electronic Systems and reorganised Barrow-in-Furness, Govan, and Scotstoun as BAE SYSTEMS Marine.

Sold in 1985, Vosper Thornycroft (VT), the other major UK builder of naval warships, also was originally part of British Shipbuilders. Over the past three decades, VT has been successful in the small-warship export market, and has been successful in building warships to destroyer size for the UK, the last being HMS *Gloucester* in 1986. To undertake a project the size of the Type 45, VT's Woolston shipyard near Southampton would have required upgrading. Rather than do this, VT is planning to move its operation into the Portsmouth Royal Naval Base and to invest in new shipbuilding facilities there. The company has signed a 125-year lease for part of the base. The lease will go into effect shortly. The Type 45 should provide a solid foundation from which the company can sustain its export business and enter competitions for future naval programmes.

Several other shipyards, including Swan Hunter, Harland and Wolff, and Appledore, have tended to rely primarily on building commercial ships but have recently entered or re-entered into competition for naval auxiliaries or warships.

To summarise the current status of the UK shipbuilding industrial base, only two UK shipbuilders—BAE SYSTEMS Marine and VT—have the combination of physical facilities, management and labour experience, and financial resources necessary to construct Type 45 ships.[2] Three other shipyards are now engaged in building ships for the MOD. Some key descriptors of these five shipyards are displayed in Table 2.1. Their locations, along with those of other UK shipyards, are shown in Figure 2.1.

ROYAL NAVY: CURRENT SNAPSHOT[3]

The Royal Navy has more than 125 ships and submarines and 182 aircraft. It employs approximately 45,000 uniformed and 20,000 civilian personnel, and operates from three major bases: Portsmouth, Devonport, and Faslane. Table 2.2 displays the composition of the fleet. There are 107 combatant ships; the remainder are auxiliary and support ships. Destroyers and frigates make up about one-third of the ships in the combatant fleet that have displacements in excess of 1,000 metric tons.

---

[2]Other shipbuilders, such as Appledore, Swan Hunter, and Harland and Wolff, could also construct the Hull, Mechanical and Electrical (HM&E) for the Type 45 if investments were made in their facilities. However, they have limited experience with combatants of the size and complexity of the Type 45.

[3]Much of the material in the remainder of this section was drawn from documents provided by the MOD and from information on the following web sites: http://www.royal-navy.mod.uk, http://www.mod.uk, and http://www.rfa.mod.uk/index2.html.

Table 2.1

British New-Construction Shipbuilders and Products[a]

Company	Shipyard	Non-MOD Products	MOD Products
Appledore Shipbuilders		Irish Navy patrol boats, Supply Vessels, Fishery Protection Vessel	Survey Ships
BAE SYSTEMS Marine	Clyde	3 Offshore Patrol Vessels (OPVs) for Royal Brunei Navy	Type 23 Frigates, various design work
		Offshore Supply Vessel, miscellaneous module work	ALSL, Landing Craft Utility, and Auxiliary Oiler
	Barrow-in-Furness		Astute submarines, LPD, and Auxiliary Oiler
			Reactivation of 4 Upholder-class submarines for sale to Canada
Harland & Wolff Holdings plc		Drill Rig, Ferries	2 RoRos
Swan Hunter		Oil Platform	2 ALSL
Vosper Thornycroft (UK) Limited		Design only for Greek Patrol Craft	Minehunter, Future OPV (FOPV)

[a]Excludes the Type 45s that are currently on order.

RAND *MR1486-T2.1*

The Royal Navy also has access to a fleet of auxiliary and supply vessels. The Royal Fleet Auxiliary (RFA) is a civilian-manned fleet owned by the MOD. Employing more than 2,000 civilians, the RFA has as its main task supplying warships of the Royal Navy at sea with fuel, food, stores, and ammunition. It also provides aviation support for the Royal Navy, together with amphibious support and secure sea transport for Army units and their equipment. The 20 ships in the RFA fleet in 2002 were built to the rules of *Lloyd's Register* (compartmentation, damage control, habitability) and also meet the standards of the Shipping Naval Acts of 1911 and of the Maritime and Coastguard Agency. Since 1994, the head of the RFA has been a flag officer coequal with other type commanders.

Figure 2.1—UK Shipyards Involved in Major MOD Shipbuilding Projects—2000

MOD NAVAL PROCUREMENT: CURRENT AND FUTURE PRODUCTION PLANS

The MOD has plans to modernise the RN/RFA along several dimensions. The focus of this study, the Type 45, will be developed amid an ambitious programme to bring a host of new-generation warships to the fleet over the next quarter century.

Table 2.2

Royal Navy Ships and Royal Fleet Auxiliary Ships: 2002

Type/Class	Number	Tonnage
Fleet		
Royal Navy		
Aircraft Carrier	3	20,600
Destroyer	11	3,560–3,880
Frigate	21	3,500–4,200
Landing Platform Dock	1	11,060
Landing Platform Helicopter	1	21,758
Minehunter	25	450–615
Patrol (Coastal)	16	49
Patrol (Ice)	1	6,500
Patrol (Offshore)	7	925–1,427
Submarine (Fleet)	12	5,000–5,208
Submarine (Trident Class)	4	15,900
Survey Ship	4	25–13,500
Royal Fleet Auxiliary		
Fleet Tankers Small	3	11,522
Support Tanker	4	40,870–49,377
Fleet Replenishment Ship	4	23,591–33,675
Aviation Training Ship	1	28,080
Landing Ship	5	5,771–8,751
Forward Repair Ship	1	10,765
Roll-on–Roll-off Vessel	2	12,350

SOURCE: Information in this table was taken from Ministry of Defence, *Performance Report 1999/2000*, Abbey Wood, England, "Annex B: Force Structure".

NOTE: Tonnage definitions vary for different types of ships. For passenger ships, the term is *gross tonnage*; for tankers and bulk cargo ships, the term is *deadweight tonnage*; and for warships, the term is *displacement tonnage*. These are defined as follows:

Gross tonnage: a measure of the total volume of enclosed spaces in the ship. The volume-to-tonnage conversion is 100 cu ft/metric ton.

Deadweight tonnage: a measure of the total volume of the ship dedicated to carrying cargo, converted to tons of seawater (35 cu ft/metric ton).

Displacement tonnage: the volume of water displaced by the hull beneath the waterline, converted to tons of seawater (35 cu ft/metric ton).

In July 1998, the MOD completed a Strategic Defence Review (SDR) that reassessed Britain's security interests and defence needs and considered how the roles, missions, and capabilities of its forces should be adjusted to meet new strategic realities.[4] For naval forces, the review suggested continuing the shift in focus away from large-scale open-ocean warfare towards a wide range of operations in littoral areas, with a premium on versatility and deployability. This direction represents a continuation of trends since the end of the Cold War.

In addition to the continuation of several procurements already planned (e.g., new Landing Platforms Dock, Astute submarines, and Auxiliary Oilers), the SDR introduced several major new programmes, including the Future Aircraft Carrier. The SDR remains the basis of general MOD policy, but the future procurement programme continues to evolve in line with the strategic environment, financial imperatives, industrial developments, and new opportunities. A more extensive naval programme is now assumed, with the addition of two new Landing Ships Logistics and a new class of Offshore Patrol Vessels.

Thus, in addition to the Type 45, ships currently on contract for manufacture, but which have yet to enter service, include the following:

- Two Landing Platforms Dock (Albion and Bulwark)
- Two Auxiliary Oilers
- Four Landing Ships Logistics
- Two survey vessels
- Three Astute submarines
- Three Offshore Patrol Vessels.

A contract for a Strategic Sealift service, which will make available six Roll-on–Roll-off (RoRo) ships as required under a Private Finance

---

[4]Secretary of State for Defence, Rt. Hon. George Robertson MP, "Strategic Defence Review: Modern Forces for the Modern World", Defence White Paper, 13 July 1998, available at www.army.mod.uk/servingsoldier/strategicdefrev.html.

Initiative (PFI) arrangement, is also expected to be signed shortly. In a PFI arrangement, the service provider can make the ships available for the generation of commercial revenue when they are not needed by the MOD, thus delivering better value for money for the taxpayer.

Under the MOD's new acquisition strategy, future programmes will be focused on capability requirements rather than on specific numbers of platforms. MOD's plans for future ship production depend on many variables, not the least of which is the evolution of the strategic environment. This caveat needs to be borne in mind where the book addresses specific ship development and production programmes now being envisioned to satisfy the MOD's future requirements. That said, at the time of this study, the MOD's future procurement planning assumptions envisioned the following:

- Two large aircraft carriers, each capable of operating up to a total of 50 fixed-wing aircraft and helicopters from all three services at one time

- Two to three more Astute submarines

- Up to 20 Future Surface Combatants (FSC; to succeed the current Frigate force).

The following discussion of current and future plans concludes with a more detailed description of the Type 45 programme, the primary subject of the present analysis.[5]

Astute-Class Submarine[6]

The MOD plans to acquire a minimum of five new submarines, known as the Astute-class submarine, to replace the Swiftsure-class nuclear-powered attack submarines (SSNs). It will take delivery of the first of these submarines in 2005.

Invitations to tender were issued in July 1994, and competitive bids were received in June 1995. The MOD identified GEC-Marconi (now

---

[5]This discussion does not include the many different smaller craft, such as the Landing Craft Utility (LCU) and Landing Craft Vehicle Personnel (LCVP), or other capabilities for which the solution might not require extensive shipbuilding work.

[6]Formerly known as the Batch 2 Trafalgar-class submarine.

BAE SYSTEMS Marine) as its preferred bidder in December of the same year. Following protracted negotiations, using the policy of No Acceptable Price/No Contract (NAPNOC), a contract for the first three submarines was placed with GEC-Marconi as the prime contractor and announced in March 1997.[7] The contract put in place the first whole-boat prime contract for UK nuclear-powered submarines. The submarines are being built at BAE SYSTEMS Marine's Barrow-in-Furness shipyard.

The prime contract is for the design, build, and initial support of three submarines. The prime contractor will undertake the support task for a total of eight submarine-years.

Future Surface Combatant

It is intended that the FSC will succeed existing Type 22 and Type 23 Frigates. Although timing and numbers are yet to be determined, it has been assumed that delivery of the first ship of this class will take place no earlier than 2012. Because the defence review emphasised that the Royal Navy needs to support expeditionary warfare and joint operations in littoral waters, the FSC is expected to have a broader, multifunction role than the primarily ASW missions that were envisioned for the latest versions of the Type 23 that it will replace. The assumption is that this warship will be operationally versatile and affordable, and capable of being deployed through life across the full spectrum of defence missions. This operational adaptability calls for ships that will be able to acquire capabilities and technologies throughout their service lives, without recourse to traditional, non-operational, major refit processes.

Future Aircraft Carrier (CVF)

The MOD is currently in the assessment phases of a programme to produce two new aircraft carriers to replace the three existing HMS *Invincible*–class aircraft carriers. These ships are currently scheduled to enter the Royal Navy inventory in 2012 and 2015.

---

[7]The NAPNOC policy requires that, for all noncompetitive procurements with an estimated value of £1M and above, an acceptable firm/fixed-price or maximum-price/target-cost incentive package be agreed on before a contract is placed.

Twice as large as the Invincible-class carrier, the CVF is expected to be among the largest warships ever built in the UK. Initial estimates are that the ships could be 300 metres (m) long, have displacement of between 40,000 and 60,000 metric tons, and able to carry up to 50 aircraft. Each vessel is expected to have a manpower requirement of about 1,200 personnel (which includes the Carrier Air Group), which is roughly equal to that of the Invincible class.

It is intended that the two ships of the new class will replace the three Invincible-class ships currently in service. This reduction is to be achieved through modern build-and-support techniques (e.g., better paints, more-durable piping) that will dispense with the need for long refit periods and will allow required availability to be achieved from only two hulls.

The CVF will be a Joint Defence Asset and will primarily be a platform for the Joint Strike Fighter (envisioned as operating in a joint RN/Royal Air Force [RAF] force), enabling operations from forces of all three services to contribute to sea, land, and air battles. Compared with Invincible-class carriers, which originally were designed for helicopter-based anti-submarine operations, the CVF will provide greater offensive air power and a larger array of aircraft able to operate in a wider range of roles.

Currently, two prime contractors are competing for the design and manufacture of the future carriers: BAE SYSTEMS and Thales (formerly Thomson CSF). Each Competing Prime Contractor (CPC) has submitted an initial Demonstration and Manufacture (D&M) plan to the MOD. The MOD intends to place an order with a single prime contractor for delivery of two vessels, the first of which is due to enter service in 2012. Because of the large size of the carriers and the limited facilities and workforce at most UK shipyards, it is envisioned that the work for the future carrier programme may be spread among several different shipbuilders.

Landing Platform Dock (Replacement) [LPD(R)]

This project covers the replacement of the amphibious assault ships HMS *Fearless* and HMS *Intrepid*, whose ages exceed 30 years. The MOD is acquiring replacements for both, which are due in service in 2003. The MOD plans to augment these purchases by acquiring eight associated specialised landing craft, four for each LPD(R), and two training craft, which successfully ran trials as prototypes in 2000. These craft are scheduled for delivery between 2001 and 2003.

The two LPD(R) are intended to provide command and control capabilities for amphibious operations. They will have two flight-deck spots for Merlin-size helicopters and will carry four Mark 10 landing craft in a floodable well deck and four Mark 5 landing craft in davits. With 550 linear metres of vehicle parking space, the vessels will have a cargo capacity of 31 main battle tanks or 16-ton trucks, 36 smaller vehicles, and 30 metric tons of stores.

The LPD(R)s are being constructed by BAE SYSTEMS Marine at its Barrow-in-Furness yard.

Future Offshore Patrol Vessels (FOPVs)

Vosper Thornycroft has been awarded a contract to lease three FOPVs to the Royal Navy and to support those vessels for five years. The vessels will be used to perform open-ocean patrols and other missions. All will be built at VT's Woolston shipyard.

Survey Vessels

The Royal Navy's Surveying Service, which has been operating throughout the world since the formation of the Hydrographic Department in 1795, is responsible for hydrographic and oceanographic surveying. From its survey data, the Royal Navy produces Admiralty charts and other nautical information used worldwide.

Two ships—HMS *Echo* and HMS *Enterprise*—are currently under construction at Appledore Shipbuilders, with VT as the prime contractor. Delivery is planned for 2002.

Type 45 Destroyer

The MOD intends to acquire up to 12 new-generation destroyers from 2007. The Type 45s will be the largest and most powerful air defence destroyers ever operated by the Royal Navy and the largest general-purpose surface warship (excluding aircraft carriers and amphibious ships) to join the fleet since designs adapted from World War II cruisers in the early 1960s.

The original intent was to buy the Common New Generation Frigate (CNGF), a collaborative programme between the United Kingdom, France, and Italy to procure a new class of anti–air warfare warship to replace the existing AAW ships. However, the trilateral ship-build part of the programme was halted in 1999, and the UK started the Type 45 project.[8] When the destroyers enter service later this decade, they will provide the fleet with an air defence capability significantly greater than that provided by the existing Type 42s.

The main armament of the class will be the Principal Anti-Air Missile System, which is being developed and procured jointly with France and Italy. PAAMS is designed to simultaneously control several supersonic, stealthy, highly maneuverable missiles that could use sea-skimming or steep-diving profiles, and each of which could engage individual targets. It will equip the Type 45 to defend itself and other ships in company from attack by existing and future anti-ship missiles, by preventing attackers from swamping the fleet's air defences. Using PAAMS, the Type 45 also will be able to operate close inshore and provide air cover to British forces engaged in land battles. The UK's share of the cost of full development and initial production of the first PAAMS will be approximately £1 billion.

In addition to a main gun for shore bombardment, the Type 45 will have either the Merlin or the Lynx helicopter, which will carry Stingray anti-submarine torpedoes. Should the requirement for a land-attack capability arise, the Type 45 is large and spacious enough to accommodate lengthened vertical launchers that could carry

---

[8]The initial Type 45 effort comprised two distinct programmes: the Principal Anti-Air Missile System (PAAMS) and the ship and its other systems (Horizon). Memoranda of Understanding (MOU) were signed by the three nations in July 1994 and March 1996. For Horizon, an initial design and validation phase (Phase 1) started in March 1996. This was to have been followed by Phase 2, the detailed design and build of three first-of-class (FOC) warships (one for each nation), to be procured under a single prime contract. For PAAMS, the next major milestone was to be the start of PAAMS Full-Scale Engineering Development and Initial Production (FSED/IP). In April 1999, ministers of the three nations announced that it was their intention to place the PAAMS FSED/IP contract quickly but that it would not be cost-effective to pursue a single prime contract for the warship. A MOD Integrated Project Team (IPT) then took work on the warship programme forward, and the Type 45 emerged.

cruise missiles. The ship also will be able to embark a force of up to 60 Royal Marine Commandos or other troops and use its aircraft and boats to support them on operations.

From 1994 to 1999, when the United Kingdom was a partner in the three-nation Horizon programme to develop a CNGF, GEC, together with VT and BAE SYSTEMS Marine, led its industry participation.[9] That collaboration ended when the Horizon programme was terminated in April 1999, but it undoubtedly provided some foundation for subsequent alliances on the Type 45 programme.

The initial contract, which was limited to initial design work and was focused specifically on the Type 45 programme, was with Marconi Electric Systems (then a part of GEC that would soon merge with BAE SYSTEMS) in November 1999.[10] In May 2000, VT and BAE SYSTEMS Marine formed an alliance to bid on the final development and construction of the Type 45. In September 2000, the Alliance submitted a bid to the MOD for sharing development and construction of the initial three ships. A contract for that work was let to the PCO in December 2000. In July 2001, the Secretary of State for Defence, Mr. Geoffrey Hoon, announced that the contract would be amended to cover the construction of the initial six ships.

Thus, by the end of 2000, it appeared that a structure and process was emerging for distributing the Type 45 business across at least three of the major shipyards (BAE SYSTEMS Marine's Barrow-in-Furness and Clyde [Scotstoun] yards, and VT's yard), with the possibility that BAE SYSTEMS Marine would distribute part of the work to its Clyde (Govan) facility. These arrangements ensured that some degree of competition would be feasible between at least VT and BAE SYSTEMS Marine for later phases of the Type 45 construction programme. They provided a promise of cost control over the life of the programme, plus the added benefit of sustaining a broad-based in-

---

[9]No documentation has been found on the exact structure of that team or the relative participation of the various members.

[10]On November 28, 1999, Marconi Electronic Systems was appointed as the prime contractor for the Type 45 programme. This responsibility was passed to BAE SYSTEMS Electronics when the merger of Marconi Electronic Systems and British Aerospace took place.

dustry that would be modernised and able to compete for future warship projects.

Those arrangements were complicated in December 2000 when BAE SYSTEMS Marine made an unsolicited proposal to build all 12 Type 45 ships (and other vessels) by itself. While apparently offering an attractive price, such an agreement would constrain opportunity for downstream competition-driven cost control on the Type 45. What effect it would have on the overall competitiveness of the UK shipbuilding industry was less apparent: While possibly strengthening BAE SYSTEMS Marine's competitive posture, it would weaken that of VT.

Thus, BAE SYSTEMS Marine's unsolicited proposal for the Type 45 programme threw into sharp focus the need for decisions on specific issues that had not previously been so clearly presented: the long-term benefits to be expected from a concentrated, monopolistic industry versus a more distributed (and, presumably, more competitive) industry.

RFA PROCUREMENTS

There are also plans to expand or modernise the RFA's tanker and cargo ships over the next 25 years; in particular, the MOD intends to take delivery of two new Auxiliary Oilers. The MOD also intends to sign a contract giving it the use of six new RoRo vessels under a PFI arrangement.

Alternative Landing Ship Logistics

The MOD is procuring four ALSL between 2002 and 2005. The ALSL will underpin amphibious operations by carrying most of the amphibious force and its equipment into theatres of operation. Whereas the LPD(R) provides the command and control function for amphibious operations, ALSL will embark the largest balance of men, vehicles, and stores needed for such operations.

The ALSL will succeed current Landing Ship Logistics (LSL) ships. Representing a marked increase in capability over the ships currently in the fleet, each ALSL will have nearly twice the carrying capacity and will be much quicker to offload.

In mid-2000, three companies bid on the contract to build the first two ALSL: Appledore Shipbuilders Ltd., BAE SYSTEMS, and Swan Hunter Ltd. Swan Hunter won that competition. As it assessed the tenders, MOD looked carefully at the strong operational reasons for replacing more of the ageing LSL and identified sufficient funding for two additional ALSL. To get the additional ships introduced into service more quickly than serial construction by one shipyard would have allowed, MOD chose Swan Hunter to design and build two ALSL, with BAE SYSTEMS Marine constructing the other two ALSL at its Clyde (Govan) shipyard, using Swan Hunter's design.

Auxiliary Oilers (AO)

The RFA plans to take delivery of two AOs in 2002. Each will have nearly 20,000 cubic metres total tankage for diesel and aviation fuel, lube oil, and cargo water; 500 cubic metres for dry cargo; and space for eight 20-foot refrigerated provision containers. Both are being built by BAE SYSTEMS Marine, one at Barrow-in-Furness, the other at Clyde (Govan).

Roll-On–Roll-Off Vessels

The Strategic Defence Review identified a need for six RoRos to provide strategic sealift to Joint Rapid Reaction Forces. Operational experience has demonstrated the difficulties in obtaining suitable ships to move military equipment in the short timescales demanded by such specialised forces and for the generalised armed forces' needs in operations worldwide.

Only major operations and exercises will require MOD to use all six ships. Therefore, the MOD has pursued a contract for a long-term service under the Private Finance Initiative.

AWSR Shipping Ltd. has been selected as the preferred bidder. The ships will be fully crewed by British merchant navy personnel while in MOD use. When combat operations require them, the seafarers will be eligible for call-out as Sponsored Reserves.

Two of the ships will be built at the Harland and Wolff shipyard in Belfast, Ireland; the other four ships will be built at the Flensburger shipyard in Germany.

PLANNED SCHEDULE OF FUTURE SHIP PRODUCTION

As noted earlier in this book, the numbers and timing of future ship programmes depend upon many variables. Table 2.3 presents illustrative MOD ship-production plans for the vessels described in the preceding section.

Table 2.3

MOD Naval Procurement: Current and Future Plans

Type/Class	Number	Tonnage	Delivery Years
Fleet			
Royal Navy			
Aircraft Carrier (CVF)	2	40,000–60,000	2012–2015
Destroyer (Type 45)	12[a]	6,500	From 2007
Future Surface Combatant	20[a]	????	Not before 2012
Landing Platform Dock	2	11,060	2003–2004
Patrol (Offshore)	3	1,700	2002–2003
Submarine (Astute)	5/6	5,000–5,208	2005–2013
Survey	2	3,500	2002
Royal Fleet Auxiliary			
Auxiliary Oiler	2	18,200	2002
Alternative Landing Ship Logistics	4	16,160	2002–2005
Other			
Roll-on–Roll-off Vessel	6	10,000	Complete by 2003

SOURCE: The information in this table was taken from Ministry of Defence, *Performance Report 1999/2000*, Abbey Wood, England, "Annex B: Force Structure".

[a]Maximum class size.

DISTRIBUTION OF SHIP CONSTRUCTION ACROSS SHIPYARDS

We have described the status of the UK shipbuilding industrial base and outlined the current and future shipbuilding programmes that have been postulated to satisfy MOD requirements. Figure 2.2 shows the distribution of present shipbuilding production activity among the shipyards, as well as an illustrative schedule for the Type 45, CVF, and FSC. How that future business should be distributed across the available shipyards is one of the critical issues facing the MOD. MOD decided, with support by this analysis, that the first six Type 45s will

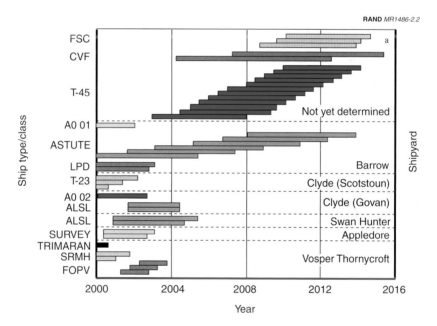

RAND MR1486-2.2

aOnly the first three ships of the FSC class are shown. There may be a total of 20 FSCs, with the start and delivery dates of the remaining 17 ships spaced approximately six months apart.

Figure 2.2—Royal Navy Warship Production Schedules by Ship Class (left) and Assigned Shipyard (right). The schedules shown are illustrative.

be built at BAE SYSTEMS Marine's Clyde (Govan and Scotstoun) and Barrow facilities and by VT at Portsmouth.

At present (2001), six shipyards have some naval shipbuilding under way. However, only one yard (Barrow-in-Furness) has an order book extending beyond the next three or four years. For the Type 45 and Future Carrier programmes to be focused on one or two shipyards would likely drive one or more of the remaining shipyards to abandon naval shipbuilding and, in turn, would tend to lessen the opportunities open to the MOD for obtaining the potential benefits of competition in future programmes. The acquisition policy and strategy options this situation presents to the MOD are explored in the next chapter.

Chapter Three

IDENTIFYING AND ANALYSING MOD'S ACQUISITION CHOICES

In managing the procurement of future naval vessels, most immediately the Type 45 programme, the MOD will face a number of issues and options. In this chapter, we describe the basic issues and options, and the analytic process used to evaluate them.

ISSUES AND OPTIONS

As noted in Chapter One, the original Alliance plan to share the workload for the first three Type 45 ships between BAE SYSTEMS Marine and VT, with competition to decide the distribution of the remaining ships in the class, was profoundly shaken by the BAE SYSTEMS Marine's unsolicited proposal to produce all 12 ships. The MOD was faced with several options for distributing the business across those firms and shipyards, each yielding a different set of advantages and disadvantages. In broad terms, the options are as follows:

- *Sole-source procurement:* In practice, the source for Type 45 production would be BAE SYSTEMS Marine, because the other potential builder, VT, does not have the capacity for the entire project.

- *Dual-source procurement:* Production would be distributed across two companies, with distribution taking place in a competitive environment, through a "directed-buy"[1] option, or in a mixture of the two.

When evaluating these options, the MOD is concerned with at least three major objectives: controlling the total cost of ship construction; providing incentives for technological innovation and the most effective use of industrial resources; and contributing to a strengthened shipbuilding industry that would be capable of effectively competing for future naval and commercial business.

In this study, we addressed these options and objectives in two ways. First, we focused on certain elements of production cost that we believed to be sensitive to an acquisition strategy for the Type 45—direct labour costs, labour transition costs, and overhead costs—rather than the total cost of producing the Type 45. Specifically, our examination excluded the cost of weapon systems that will be placed on the Type 45. Production costs for the Type 45 should not be treated in isolation. We considered the total work from all programmes at a shipyard.

Second, having investigated these costs, we subjectively examined how competition might affect other objectives for the Type 45 programme, for other MOD shipbuilding programmes, and for the overall health and performance of the UK shipbuilding industrial base.

QUANTITATIVE EFFECT OF USING TWO SHIPYARDS

How the workload for the 12 ships in the Type 45 programme is distributed between the two shipbuilders affects not only the total labour hours needed to construct the ships but also how those total labour hours are spread among the four shipyards.

In the absence of any workload reductions resulting from competition, sole-source procurement will theoretically result in the lowest number of total labour hours for the programme. This is a result of the learning effect, a widely accepted principle in manufacturing industries. According to the principle, the workforce becomes more

---

[1]*Directed buy* entails the buyer defining and directing the distribution of work among firms, rather than unfettered competition determining the distribution.

proficient when activities are done on a repetitive basis, which implies that fewer hours will be required to build each successive ship. A learning curve describes how the hours to build each ship decrease with successive ships. For example, a 90-percent learning curve suggests that doubling the number of ships built by a workforce would reduce the labour hours to construct the ships by 10 percent. Therefore, if the first ship required, say, 1 million man-hours to build, the second ship would need only 900,000 hours, and the fourth ship would require only 810,000 man-hours.[2]

Because of this learning effect, and temporarily ignoring any possible offsetting effects of competition, distributing the 12 Type 45s between two producers would most likely result in an increase in the total hours over those for all units being built by one producer. Figure 3.1 shows the theoretical increase in the total man-hours for the 12 ships for different distributions between the producers and for different learning-curve slopes. For example, if the 12 ships were split equally between the two producers (the worst distribution in terms of total man-hours), the total man-hours would increase by approximately 9 percent for a 90-percent learning curve and by almost 19 percent for an 80-percent learning curve.

Two important assumptions are inherent in this two-producer-induced increase in man-hours, and they lead to two important questions. The theoretical calculation behind Figure 3.1 assumes that the learning curves at the two producers are the same, thus making the curves symmetrical around the 6/6 split. If one manufacturer has greater or lesser learning than the other, the curves would not be symmetrical and would show a different increase in man-hours.

_____

[2]This example follows the "unit cost" theory (as opposed to the "cumulative average cost" theory), for which the average cost of the first two units would be 90 percent of the first unit and the average cost of the first four units would be 81 percent of the first unit. The unit theory is generally accepted as appropriate for major weapon-system production.

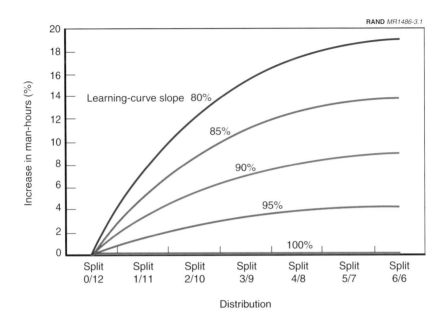

**Figure 3.1—Increase in Total Man-Hours for Different Symmetrical
Learning Curves and Distributions of the 12 Type 45 Ships**

Second, although the theoretical increases shown in Figure 3.1 may
be appropriate for a directed-buy option, they may not be correct for
the competition case. In theory, competing firms find ways to
increase learning from ship to ship, or they find ways to reduce the
time to build a ship in order to win subsequent bids. In fact, the expectation is that competition will potentially overcome the increase
in man-hours shown in Figure 3.1 (and other cost effects) and,
therefore, lead to lower costs than those resulting from the sole-
source option.

This leads to two important questions: *Is learning greater for competitive programmes than for noncompetitive programmes? What is
the appropriate learning curve to use?*

To address these questions, we gathered man-hour data from 25 different United Kingdom and U.S. Navy ship programmes. Table 3.1 summarises our statistical analysis of these programmes.[3]

The analysis indicates that the average learning curve for the 25 programmes was 87 percent and that the learning curve for competitive programmes was slightly better (2 percentage points) than that for noncompetitive programmes, although not a significant difference.[4] Figure 3.2 shows the histograms for the complete sample of competitive and noncompetitive programmes.

Table 3.1

Summary of Ship-Production Learning-Curve Statistics

Summary	Average Slope (percent)	Standard Deviation	Sample Size
All programmes	87	5.2	25
All competitive programmes	86	5.9	14
All noncompetitive programmes	88	4.6	13
UK-military competitive programmes	83	5.3	3
UK-military noncompetitive programmes	86	4.9	5

NOTE: Data for individual programmes are privileged and thus can be displayed only in a summary format.

RAND *MR1486-T3.1*

---

[3]This analysis was prepared by Fred Timson at RAND.

[4]The standard deviations are large and the sample sizes are small; therefore, it is not obvious from casual inspection that the differences between the competitive and noncompetitive samples are meaningful. To formally examine the difference, a statistical test is run to determine whether differences between means are significant. One such test, a "t-test", determines the likelihood that the difference is due to random chance. If that likelihood is low (usually ≤ 5 percent), the difference between the mean values can be assumed to be meaningful—i.e., the difference is "real". The generally accepted levels of significance are 5 or 1 percent. Occasionally, a result may be described as significant at the 10-percent level. In our particular case, the likelihood is

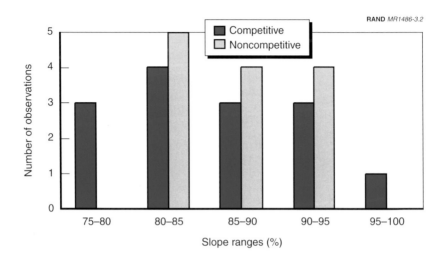

**Figure 3.2—Frequency Distribution of Learning-Curve Slopes for
Competitive and Noncompetitive Programmes**

The average learning curve of 87 percent suggests that, theoretically,
splitting the 12 ships between the two producers may increase man-
hours by as much as 11 percent.

OUR ANALYTIC APPROACH

Drawing from and building on previous RAND studies that examined
the economic effects of various procurement strategies (Birkler et al.,
1994, 1998), we developed an analytic model of the UK shipbuilding
industrial base specifically for the Type 45 programme. The model
takes into consideration the many unique aspects of the programme.
It also considers all current and future programmes at BAE SYSTEMS

---

much greater than 5 percent, indicating that the difference could be entirely due to
chance (incomplete sampling). For the competitive and noncompetitive subsets of all
the ship programmes considered, the likelihood of concluding that the two sets are
different when they are really the same is approximately 30 percent; the t-statistic is
1.05 with 26 degrees of freedom (= t – sample size). For the UK military programmes,
the likelihood is about 40 percent, and the t-statistic is 0.89 with 5 degrees of freedom.
Thus, we cannot say that there is a statistically significant difference between
competitive and noncompetitive programmes.

Marine and VT shipyards and calculates the workforce, overhead, and investment costs. The results were validated with contractors and various MOD offices.

Using the option of having BAE SYSTEMS Marine build all 12 Type 45s as the baseline—its unsolicited proposal—the model displays the relative cost effects on the Type 45 and other programmes for alternative procurement paths. However, as we described in the preceding section on learning curves, we do not know how competition will affect the learning curves at the shipyards or the number of hours to build successive ships. No one has been able to adequately quantify how much competition saves in weapons procurement— how much each competitor is willing to invest to reduce costs, how much risk each is willing to assume, or how efficient each competitor can become. In our model, we estimate only the costs of specific distributions of the total workload without taking into account any effect of competition. The results of these calculations show the incremental costs of distributing production among two or more producers compared with the costs of a single producer.

Shipyard Labour Model

Over the past several years, RAND has developed analytic methods to examine workforce issues at shipyards. The total labour cost for all the work at a shipyard is more than the sum of its individual components. Labour costs at shipyards have nonlinear aspects whereby a change in an input to the system does not result in a proportional change in an output (cost). For example, if we increase the number of ships that we build by 20 percent, the total production cost does not increase by 20 percent; it increases by something less than 20 percent. Similarly, a decrease of 20 percent in the number of ships produced results in a production-cost savings that is less than 20 percent. RAND's research has focused on modelling these nonlinear aspects of labour cost.

Many factors cause this nonlinearity. Overhead and burden are one example. *Overhead costs* are those costs that are related to production activities but that cannot be allocated directly to a particular product because of either practicality or accounting convention. Overhead includes the costs of fringe benefits, indirect labour, depreciation, building maintenance and insurance, computer services, supplies, travel, and so forth (Cash, 1999). *Burden* is the sum of overheads, G&A (General and Administrative), and profits. The overhead rate is generally *inversely* proportional to the total workload. So, removing work from a shipyard lowers the direct labour cost but increases the indirect rate for the remaining work. Another nonlinear component involves worker productivity and training cost.

Ideally, a production facility has a workload that is fairly stable over time or that grows moderately. When workload changes rapidly or fluctuates, additional costs are introduced. For example, a facility would have to hire many new workers if workload were to increase rapidly, and it is likely that many of these new workers would be inexperienced. Hence, productivity would suffer. In a situation of increasing workload, therefore, we would expect the workforce to have lower overall productivity, causing labour costs to rise. Additional direct hours must be worked to complete the tasks, and there would be some nonrecurring fixed cost to train these new workers.

The following is a partial list of nonlinear factors in shipbuilding:

- *Training Costs*: costs associated with training new employees, whether in direct cash outlays, such as external courses, or nonproductive time.

- *Hiring Costs*: costs associated with hiring new employees, which may be recruitment fees, bonuses, management time, etc.

- *Termination Costs*: compensation that may be owed to a worker upon termination of employment. This compensation is typically some fraction of his or her annual salary.

- *Productivity Costs*: higher labour costs to cover the extra hours that new workers require to perform a specific task.[5] As described above, new workers hired by a shipyard may be inexperienced and, therefore, take longer to complete tasks than it takes experienced workers. Over time, new hires gain experience,

---

[5]For our example, we have assumed that the hourly wage rates are the same. In the United States, the wage rate of a new worker is usually lower than that of an experienced worker. This wage differential somewhat compensates for the lower productivity.

lessening the learning effect and increasing productivity. However, the effect of dramatically increasing the workforce at a facility can be seen for several years.

- *Constraints in Ability to Expand or Contract*: limits placed on a shipyard by the available labour pool, mentoring ratios, etc. A shipyard does not have an infinite labour pool upon which to draw new workers, so there might be a maximum rate at which a yard could expand. Even when there may be available workers, there could also be some limit on how many new employees a yard can accommodate at one time (training and mentoring ratios may limit this rate). These constraints would cause the shipyard to carry more people than necessary at times in order to meet peak demands in the future, a situation that, again, would tend to increase total labour cost.

- *Learning*: proficiency that occurs when building multiple ships of the same class. A workforce becomes more proficient with each successive ship; hence, the labour hours and costs decrease. This learning effect has important implications when we look at sole-source versus dual-source acquisition strategies. With all other things being equal (and ignoring competitive pressures), the cost to build a class of ships should take fewer hours if all the work is concentrated at a single producer.

- *Overhead and Burden Rates*: costs of fringe benefits, indirect labour, depreciation, building maintenance and insurance, computer services, supplies, travel, etc., and the sum of overhead, G&A, and profit. Overhead rates are inversely proportional to workload. When workload decreases, overhead rates increase, reducing any savings from removing the work.

Methodology

The goal of the model is to estimate the labour costs, overhead rates, labour-force transition costs, and learning improvements across several shipyards for a given shipbuilding strategy.

Once the acquisition plan was determined, we calculated the labour demands at each shipyard. Every shipyard project that might be built carries a labour profile that shows the man-hours, or equivalent workers, per quarter over the build period. These labour profiles are unique to each shipyard.[6]   After applying an appropriate learning factor, we summed these labour profiles across all programmes at the shipyard to determine the total labour demand at a yard. This total labour is what we term *required labour:* the minimum labour necessary to complete all tasks at each shipyard.

However, the required-labour level may not be the actual level that the shipyard would employ. Hiring and firing constraints (e.g., mentoring ratios) may cause the shipyard to hire or retain more workers than are absolutely required. Given labour constraints that restrict the rate of workforce expansion from one period to the next, we used a linear programming model to determine the effective labour. *Effective labour* is always equal to or greater than the required labour.

From the effective labour, we established an overall burden rate for the shipyard. We modelled burden as having a fixed component and a variable component. The burden rate equation for a shipyard takes the following form:

$$\text{Burden}_{\text{quarter}} = \frac{A}{\sum\limits_{\text{trade}} \text{hours}_{\text{quarter, trade}}} + B \qquad (3.1)$$

where A and B are constants specific to each shipyard. Note that as the total site hours increase, the burden rate decreases. This elastic model works well if there are not large, long-term changes in employment levels (particularly, downward ones). In practice, a shipyard would reduce its fixed burden costs in such an environment, possibly by shedding unneeded facilities and/or reducing the indirect staff and support. To reflect such a situation in the model, we applied a cap (200 percent) to the burden rate.

Lastly, after determining both the effective hours and burden rates, we calculated total labour cost, as follows:

---

[6]Labour profiles are developed for three trades—direct labour, support, and engineering. BAE SYSTEMS Marine and VT provided the build periods, workload profiles, and learning-curve assumptions.

$$\text{Cost}_{\text{quarter}} = \left(1 + \text{Burden}_{\text{quarter}}\right) \times \sum_{\text{trade}} \text{rate}_{\text{trade}} \times \text{hours}_{\text{quarter, trade}} \quad (3.2)$$

where

> $\text{rate}_{\text{trade}}$ is the wage rate for a specific labour type.

Data

Obviously, such a model requires an extensive amount of data about each shipyard and ship class. RAND researchers prepared comprehensive data-collection forms so that such data would be collected in a consistent fashion. The types of data gathered were as follows:

- *Shipyard Capacity*: steel throughput, docks, lifting capacity, outfitting berths, etc.

- *Workforce Profile*: experience levels, age, productivity, costs for hiring and training, termination costs, restrictions on hiring and termination, current employment levels, and the use of contract workers.

- *Production Experience*: numbers and types of ships built over the past five years (including commercial work).

- *Current and Future Production*: current and anticipated production plans (by ship).

- *Workload Projections*: for each activity listed in the current and future production plan, a listing of the labour profile, by trade and quarter. Further, these data include design and development workload.

- *Wage Rates*: hourly wage rates for all the labour types.

- *Burden Rates*: overheads (OH), G&A, and profit rates as a function of different sites' workloads.

- *Investment Levels*: fixed investments, such as facilities, necessary for a particular programme or investment required overall.

We requested this information from five shipbuilders: VT, BAE SYSTEMS Marine, Swan Hunter, Appledore, and Harland and Wolff. We also held discussions during site visits with each firm. Beyond the shipbuilders, RAND had similar discussions with Type 45 and Astute PCOs, as well as with all current MOD ship programme managers and their staffs. These offices provided considerable supporting information and data about their ship programmes.

Specific Assumptions

To use the quantitative model to estimate the costs for the various Type 45 acquisition options, we made several simplifying assumptions (see Appendix A for the results of sensitivity analyses involving these assumptions). They include:

- *Shipyards*: For this initial analysis, we considered only BAE SYSTEMS Marine shipyards (Barrow-in-Furness and Clyde) and the VT shipyard (assuming it relocated to Portsmouth).

- *Time Period*: 2001 to 2014. This period covers the planned construction of the Type 45s.

- *Shipbuilder-Provided Data*:   site-specific learning curves, labour-force data, and overhead rates.

- *Workload Allocation*: for the FSC, split evenly between BAE SYSTEMS Marine and VT; for the CVF, one-half to BAE SYSTEMS Marine and one-quarter to VT.[7]

- *Maximum Overhead Rates*:   200 percent.

- *Facility Investment*: For the two shipyards (VT and Barrow-in-Furness) to compete for the Type 45 production, each site would require some facility investment/improvement. Given an early, order-of-magnitude estimate by the Type 45 PCO, we used the following assumptions in our calculations:

  — Barrow-in-Furness—The investment would depend on the extent of the improvements required. We assumed that this investment cost at Barrow-in-Furness would be independent of the number of ships BAE SYSTEMS Marine won. In other words, this shipyard would build facilities to the full rate of

---

[7]This is RAND's allocation and does *not* reflect any MOD input or guidance.

production, not knowing how much of the production it would win; so, its investment costs would be roughly the same under sole-source and competition.

— VT—These costs are more difficult to determine,[8] because there was no detailed estimate on the cost to relocate VT to Portsmouth. As another rough estimate, the PCO thought that nearly £2.5 million would be needed to relocate VT to Portsmouth and another £25 to £50 million would be required to upgrade the Portsmouth site to produce the Type 45. Given the lack of more detailed estimates by VT, and erring on the conservative side, we assumed that the site investment cost for VT would be £40 million (again, independent of production rate). For block production, we assumed that the investment cost would be half that value, or £20 million.

Break-Even Approach

Competition is just one form of a multiple-source production strategy. Other strategies can allocate production among several firms. Unfortunately, there is no reliable method for predicting the savings from multiple-source procurement (competitive or otherwise). We cannot predict the behaviour of the firms or their willingness to reduce costs and profits in order to undercut the other firm or firms. *Thus, the basic question is not how much money will be saved but, rather, whether introducing an additional production source is a reasonable strategy to pursue.*

One way to make that reasonableness determination is through a break-even analysis (Margolis, Bonesteele, and Wilson, 1985; Hampton, 1984). Such an analysis does not require an explicit estimate of the savings from multiple-source production; rather, it deduces the magnitude of savings needed to justify a second source of production. In general terms, "break-even" refers to the point at which the expected cost to the government of a multiple-source al-

---

[8]The estimates were made by RAND researchers and do not reflect VT's actual figures.

ternative equals the cost of the sole-source alternative, which, for the Type 45, is the unsolicited proposal from BAE SYSTEMS Marine.

To calculate a break-even value, we used our shipyard model to determine the production costs for various strategies. Our metric is a percent change in production cost relative to the BAE SYSTEMS Marine's sole-source case. We include in this percent not only the effect on the Type 45 programme but also the effect on other contemporaneous programmes. It is a net cost delta for the UK government:

$$\text{Break - even percent} = \frac{\sum_{\text{All Programs}} \text{Cost}_{\text{Dual Source}} - \sum_{\text{All Programs}} \text{Cost}_{\text{Sole Source}}}{\sum_{\text{Type 45}} \text{Cost}_{\text{Dual Source}}}$$

$$(3.3)$$

For competitive strategies, we can compare this break-even value with historical values to assess whether competition could reasonably lead to overall cost savings. We make such comparisons in the next chapter.

Chapter Four

ANALYSIS OF PROCUREMENT ALTERNATIVES

Various acquisition strategies are open to the MOD for the Type 45. In this chapter, we first describe the five alternative procurement strategies examined in this study, then show the expected labour costs of each. Next, we discuss the factors other than labour cost—risks and rewards—that might be affected by a competitive strategy. We conclude this discussion with an overall integration of the analysis and observations on all the material in the chapter.

CASES EXAMINED IN THE COST ANALYSIS

A central objective of this study was to examine the cost consequences of different strategies for producing the Type 45. We examined five basic strategies on three procurement paths:

1. **Sole-source procurement.** All production is performed by a single firm.

2. **Dual-source competitive procurement.** Production is split between two firms, with the split determined by head-to-head cost competition. Two variations were examined:

    2a. Competitive procurement of whole ships

    2b. Alliance proposal, involving assigned production of the first three ships (with the first-of-class built in blocks at multiple shipyards) and competitive procurement of the remainder.

3. **Dual-source, directed buy.** Production is split between two firms, but the split is directed by the MOD, without head-to-head price competition extending throughout the life of the programme. Two variations were examined:

   3a.   Assigned production of whole ships

   3b.   Assigned production of individual ship components (blocks).

COMPARISON OF LABOUR COSTS ACROSS STRATEGIES

As described in the preceding chapter, we calculated costs using identical assumptions for both competitive and noncompetitive strategies. That is, we made no attempt to adjust costs to reflect the fact that firms might perform differently according to the degree of competition present. Our objective here is to estimate the direct consequences of spreading the production between two firms rather than concentrating it at one firm. That cost difference represents the amount of cost reduction that would have to be created through competitive pressure in order for the MOD to break even by sustaining a competitive environment for Type 45 production, or that would have to be justified on some other criterion if a dual-source strategy were adopted.

The Reference Case: Sole-Source Production

Our analysis approach entails a basic assumption: Any procurement strategy that concentrates all production at one firm cannot be truly competitive. A single-source strategy is likely to entail fierce competitive bidding at the start of the construction programme; however, as soon as an award is made to a single firm, the competitive environment disappears and that firm becomes the monopolist supplier for the remainder of the construction programme. If that monopolist supplier fails to perform according to the terms of the contract or is not otherwise responsive to the needs and preferences of the buyer, the buyer has limited recourse to effective remedies. The history of military systems acquisition is replete with examples of sole-source suppliers who failed to satisfy the terms of the original contract and the severe problems those failures caused the procuring agency.

For these reasons, we define a *sole-source acquisition* as a *noncompetitive strategy.*[1] We calculated the production cost for the sole-source strategy by assuming the plan described in the BAE SYSTEMS Marine unsolicited proposal, whereby all ships would be built in BAE SYSTEMS Marine facilities, thus achieving maximum learning effects as well as sharing overhead costs with other shipbuilding programmes scheduled for those facilities. The result becomes the reference point against which all dual-source options are compared.

Dual-Source, Competitive Procurement of Whole Ships

We assumed that procurement is divided into four sequential lots of three ships each, to sustain competition throughout the construction programme. A price competition would be held for each lot, and the winner would be awarded production of two ships and the loser given the option of building one ship at the winner's price.

Under this strategy, the final allocation of ships depends on which shipbuilder wins each of the three-ship competitions and whether the loser decides to build one of the three ships at the winner's price. To pare the multitude of options to a manageable few, we decided on the following three cases for the competitive options:

- Eight ships to BAE SYSTEMS Marine and four to VT

- Six ships to BAE SYSTEMS Marine and six to VT

- Four ships to BAE SYSTEMS Marine and eight to VT.

These three specific cases should provide sufficient insights to understand the relative cost differences between the sole-source option and the directed-buy and competition options.

---

[1]We recognise that the BAE SYSTEMS Marine's unsolicited offer was made in a competitive environment and offered attractive terms. However, under that proposal, no continuing competition would have existed throughout the production phase. Experience shows that, under such conditions, production costs tend to rise and the programme becomes the equivalent of a sole-source negotiated contract.

Dual-Source, Competitive Procurement: Alliance Proposal

This strategy is basically the initial plan for the Type 45 (before the unsolicited proposal from BAE SYSTEMS Marine), whereby each shipbuilder would be allocated approximately half of the total workload for the first three ships and then compete for the following three lots of three ships each, with the loser having the option of building one of the three ships at the winner's price. Again, there are a number of potential outcomes, depending on which shipbuilder wins each of the three competitions and what decision the loser makes. For simplicity, we limit the Alliance option to the case where BAE SYSTEMS Marine builds 7.5 ships and VT builds 4.5 ships.

The outcome of this case is similar to the third case in the preceding strategy whereby BAE SYSTEMS Marine builds eight ships and VT builds four (only one-half a ship difference), except that here competitive pressure is applied only to the last nine ships. Therefore, comparing the costs of the specific Alliance case to the case of eight/four should provide sufficient insights of the other potential outcomes of the Alliance competition (for example, BAE SYSTEMS Marine building 6.5 ships and VT building 5.5 ships).

Comparison of Dual-Source, Competitive Procurement with Sole-Source Procurement

Figure 4.1 shows the relative percent cost increase of the three specific whole-ship cases and the Alliance case compared with the sole-source strategy.[2] In percent terms, these increases represent how much competition would have to save in order for the Type 45 cost

---

[2]When estimating the percent increase in cost due to distributing production between two producers, we needed to include the same cost elements as were used in the historical database of competitive and noncompetitive ship production. Unfortunately, we do not know the exact array of cost elements included in the historical data, and they probably differ somewhat from programme to programme. We do know that almost all ship procurements include some Government-Furnished Equipment (GFE) that is probably not included in the stated total cost of the ship, but the stated costs shown in historical records will always include more than labour and overhead. Therefore, when calculating the percent increase in cost due to distributing Type 45 production between two producers, we added the cost of material, about £50 million, which we assume, because it is purchased by the PCO and provided to the shipyards, is the same across all options.

with competition to equal the sole-source cost. Note that the Alliance case involves a directed buy of the first three ships and competition for the last nine ships. Therefore, the cost of the competed nine ships would have to be reduced about 13 percent in order to match the cost of the sole-source option. The other three cases involve competition for all 12 ships. Therefore, for the MOD to break even, competition must yield greater percentage gains in the Alliance case for the nine competed ships than for the other three cases (where there is competition for 12 ships).

Figure 4.1 shows that, of the three competitive procurement cases, the cost relative to the sole-source option increases as VT builds more of the 12 ships. The highest increase in cost is approximately 13 percent when BAE SYSTEMS Marine builds four ships.

Figure 4.2 provides further details on what contributes to the cost increases over the sole-source option. It categorises cost by Type 45 labour and overhead, initial investment cost, Astute overhead, transportation (of the blocks from VT to Clyde [Scotstoun] in the Alliance case), and other (a decrease in the costs of other programmes at VT resulting from increased workload compared with the VT workload under the sole-source option). The figure shows that the Type 45

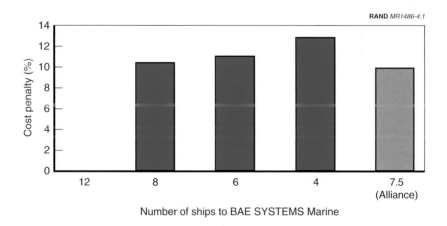

Figure 4.1—Cost Penalties of Five Whole-Ship Procurement Options

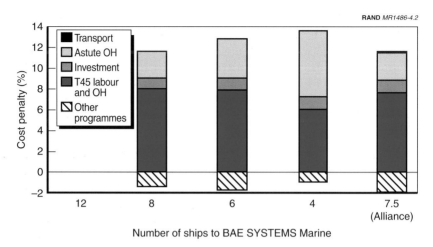

NOTE: For the whole-ship options, the cost of transportation is so small relative to the other costs that it does not show.

Figure 4.2—Composition of Cost Penalties of Five Whole-Ship Procurement Options

labour and overhead costs at first increase as BAE SYSTEMS Marine moves from building 12 ships (the sole-source option) to an even split of the ships between BAE SYSTEMS Marine and VT, then decreases as VT builds more ships (owing to the symmetrical nature of the learning-curve assumptions as described in the preceding chapter). Also, the Astute overhead increases as BAE SYSTEMS Marine builds fewer ships.

In addition to providing the cost implications of different acquisition strategies, the model also provides the size of the labour force required at the various shipyards for the different programmes ongoing at the shipyard.

Break-Even Analysis for Competition

From the information in Figures 4.1 and 4.2, we can estimate how much competition would have to lower costs compared with the costs of the sole-source option. The question is, Are these percent reductions reasonable to expect from competitive programmes? As described in Chapter Three, we gathered data from a number of historical programmes that involved competition to understand the potential effect of competition on cost. For comparative consistency, we desired historical data on ship-construction programmes that involved competition. To provide a basis for estimating total costs if conducted sole-source, we needed programmes that had an initial, noncompetitive phase, followed by a competitive phase (Birkler et al., 2001).

We found five such programmes: TAO-187, LCAC, Type 23, CG-47, and LSD-41. In this small sample, three of the five programmes resulted in savings of more than 10 percent and two of the five programmes resulted in savings of 20 percent or more. This small, ship-only dataset suggests that there is about an equal chance that competition in the Type 45 programme can lead to costs lower than those of sole-source acquisition as that the costs of competition could be greater than sole-source.

Because of the small sample size of ship programmes, we augmented our database of competitive programmes by adding missile programmes that were competitive. The distribution of percent savings for the new database of 31 programmes is shown in Figure 4.3.

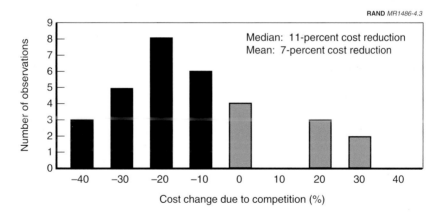

Figure 4.3—Cost Changes Resulting from Competition in 31 Missile and Ship Programmes

The figure shows that the average savings from competitive ship and missile programmes was approximately 7 percent, with a median savings of approximately 11 percent. Most competitive programmes resulted in decreased costs; however, a few programmes actually resulted in increased costs of up to 30 percent.

Given that competition in the Type 45 programme would have to reduce costs (compared to the sole-source option) from 10 to 13 percent, depending on the distribution of ships to the two shipbuilders (see Figure 4.1), the historical competitive programme data suggest that there is approximately a 50:50 chance that competition, if it can be sustained over the programme as a whole, will result in equal or lower costs than the sole source. Therefore, there is no dominant answer to whether competition or sole source would likely lead to lower costs.[3]

Dual-Source, Directed Buy of Whole Ships

One of the important potential advantages of the competitive-buy option is that two shipbuilding firms would remain in the military shipbuilding business, thus encouraging innovation, supporting improvement in facilities, and generally strengthening the UK shipbuilding industry. However, the proposed "Alliance" mechanism involves contracting for four successive blocks of three ships each, with the winning bidder being awarded two ships but with the loser having the option of building one ship at the winner's price—an arrangement offering a distinct possibility that the loser would simply exit the programme, leaving a sole source to build the remaining ships.

If the MOD wishes to ensure keeping two firms in the military shipbuilding business, one option would be to simply direct some ships to one firm and the remainder to the other firm. While forgoing the possible benefits of competition, such a strategy might be attractive for a variety of reasons other than cost minimisation, such as retain-

---

[3]Here, we focused on the effect of competition or sole source on Type 45 costs. There may also be implications for future programmes, especially if the sole-source option results in BAE SYSTEMS Marine becoming the sole warship builder in the United Kingdom. Appendix B provides an initial analysis of the potential costs of future programmes in such a monopoly situation.

ing the option of competition in future programmes and stabilising employment levels at select shipyards. However, there would likely be a cost penalty compared with sole-source procurement. Barring the introduction of other considerations, that cost penalty would be as shown for the three split-buy options displayed in Figure 4.1: roughly 10 to 13 percent.

Dual-Source, Directed Buy of Ship Blocks

In addition to acquisition options that involve the allocation of whole ships between the two shipbuilders, we examined two options that involve the directed allocation of major portions, or blocks, of the ships as depicted in Figure 4.4.[4]

Such a strategy would keep both shipbuilders involved in the Type 45 programme, and in future warship programmes, while overcoming the disadvantages, mentioned in the preceding chapter, of increased man-hours due to lower learning gains. By having each shipyard build the same blocks for all 12 ships, the maximum gain to learning can be achieved. Therefore, acquisition strategies that involve building blocks at shipyards have a number of advantages over the options of building whole ships.[5]

We examined two specific block options:

- Option I:
    - Blocks B, C, and final assembly at Barrow-in-Furness
    - Blocks D and E at VT
    - Blocks A, F, and mast sections on the Clyde.

---

[4]Since the original plans called for the first-of-class to be built in blocks at multiple shipyards and the whole ship assembled at Clyde (Scotstoun), the block concept was incorporated into the Type 45 design from the very beginning.

[5]In addition to advantages of building the ship in blocks, there are disadvantages, such as increased costs for transporting the blocks to the final assembly area. Appendix C describes historical experiences of building ships in blocks at different shipyards and provides an assessment of the potential advantages and disadvantages.

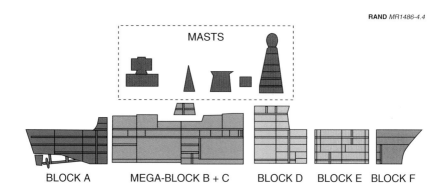

RAND *MR1486-4.4*

BLOCK A   MEGA-BLOCK B + C   BLOCK D   BLOCK E   BLOCK F

Figure 4.4—Type 45 Blocks

- Option II:
  — Blocks B, C, and final assembly at Barrow-in-Furness
  — Blocks D, E, F, and mast sections at VT
  — Block A on the Clyde.

For each of these options, we assumed that the initial investment would be £20 million less than the investment costs for the sole-source option (since VT would require less building/upgrading of facilities at Portsmouth, and Barrow-in-Furness would require the same degree of facility building/upgrading as under the sole-source option) and that the cost to transport the blocks from VT or the Clyde to Barrow-in-Furness would be £300,000 per shipment.

Figures 4.5 and 4.6 show the cost penalties for these two block options when compared with the sole-source acquisition strategy. For comparison purposes, we also show in the two figures the cost of the whole option, whereby BAE SYSTEMS Marine builds eight ships and VT builds four (the lowest-cost penalty of the whole-ship options).

As Figures 4.5 and 4.6 suggest, both block options significantly reduce the cost penalty of splitting production between two firms rather than taking the whole-ship options. This reduction results primarily from the increased learning effect of having each shipyard build all 12 of the specific blocks. However, this general strategy

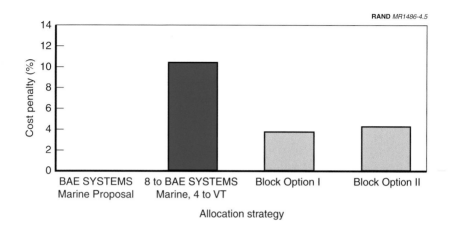

Figure 4.5—Cost Penalties Associated with Four Procurement Options

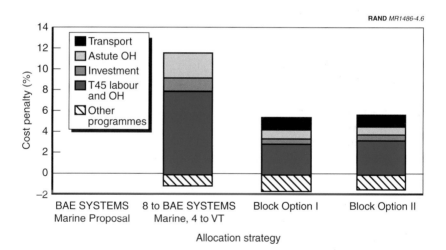

**Figure 4.6—Composition of Cost Penalties Associated with Four
Procurement Options**

provides no opportunity for continuing competition throughout the production programme, no basis for expecting any reduction from the costs shown, and minimal incentive for innovation. Therefore, this strategy must be evaluated on criteria other than that of expecting to break even on costs within the Type 45 programme.

Summary of Cost Analysis

In Figure 4.7, we summarise the estimates of production cost for the various options relative to the sole-source option.

The figure presents three dual-source strategies. The first such option, represented by the three bars on the left (dark blue), is competitive procurement of whole ships. It is expected that competitive pressures inherent in this strategy would reduce the actual costs, and our analysis suggests that there is roughly a 50:50 chance that competitive pressure would overcome the cost penalties of from 10 to over 12 percent shown in the figure. However, such cost reductions are not ensured, and the split of production between two competitors could result in a cost increase over the sole-source option.

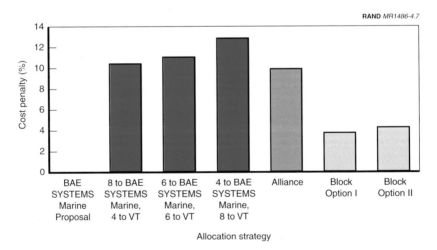

Figure 4.7—Summary Comparison of Cost Penalties of Seven Procurement Options

The second procurement option, in which the first three ships would be a directed split between the two firms, with the remaining nine ships competed, is represented by the fourth bar from the left (medium blue). The bar reflects the estimated cost penalty for the Alliance strategy. If it is expected that the cost penalty of about 10 percent is to be recovered through competition, then a cost reduction of about 13 percent would be required across the nine competed ships.

The third procurement option, wherein each firm produces a specified set of blocks for the entire 12 ships, provides no opportunity for competitive recovery of the cost penalty. However, the penalty is smaller than those for the other options and might be considered a reasonable cost to pay to obtain other possible benefits of that option.

OTHER FACTORS AFFECTING THE TYPE 45 PROCUREMENT DECISION

The analysis and results described above focused entirely on how the procurement strategy might affect the production cost of the Type 45. However, the choice from among procurement strategies will have other effects, some within the Type 45 programme and some outside the programme. Our analysis of these other likely effects—risks and rewards—has been less extensive than that for Type 45 production costs, but it provides some additional insights to support a decision on the overall Type 45 procurement strategy.

Risks and Rewards to the Type 45 Programme

The main advantage to the Type 45 programme of the sole-source option is that the costs are within the range deemed "affordable" by the MOD. The lower estimated costs of the sole-source option are primarily due to the economies of scale of having all the ships built by a single shipbuilder. As discussed earlier in this chapter, any competitive strategy would involve splitting production between two shipyards, leading to a directly estimated cost greater than for a single source, but with the expectation that competition would result in some reduction of those costs. Our calculations show that there is an even chance that competitive pressures would lead to costs no greater than the sole-source option. However, such an outcome is *not* ensured. Thus, selection of a competitive, dual-source strategy is an attractive option, but not one without cost risks. If competition yields little or no cost savings, programme execution would be more difficult within the projected budget.

The option of having specified shipyards build the entire production run of some blocks should lead to costs that are slightly higher than the sole-source option but lower than the estimated cost of competitive dual sources for the entire ship, assuming little or no cost reduction through competitive pressures. The cost of a directed buy of blocks from two shipyards falls within the range deemed affordable by the MOD, retains the advantages of sustaining two shipyards, and is not critically dependent on subsequent, and uncertain, reductions of cost through competitive pressures.

Several other factors must also be considered when deciding on an acquisition strategy for the Type 45 programme. These are discussed next.

Innovation. With a guarantee of all 12 ships, the shipbuilder may have little incentive for innovation or to find ways to reduce costs. This situation is also true for the block options, since each shipbuilder is guaranteed specific portions of the total programme. Directed buys may influence the shipbuilders to be innovative in reducing costs if they perceive the opportunity to gain a larger share of the remaining ships in the programme. Competition should foster the greatest degree of innovation as each shipbuilder strives to attain more-efficient building techniques that will result in lower costs and, it is hoped, larger portions of the total programme.

Multiple Sources. The sole-source option leaves little or no alternative if the sole shipbuilder has problems: There is no other shipbuilder in the programme to turn to. The block options improve the situation somewhat, because at least two shipbuilders are involved in the programme. However, each shipbuilder would have to "learn" and become proficient in building the blocks that were assigned to the other shipbuilder. Competitive environments and the directed-buy option result in both shipbuilders being capable of building the entire ship. These options provide the highest assurance to the MOD that the programme can continue with little disruption if one of the shipbuilders experiences problems that prevent it from building the ships in a timely and efficient manner.

Leverage on the Shipyards. Under the directed-buy option, the MOD and the Type 45 PCO may have little leverage over the shipbuilder once it knows that it will have no competition in the programme. The leverage over the shipbuilder increases with both the directed-buy option and the block option. With both shipbuilders involved in the programme, even when the total programme is allocated in some way, the MOD and PCO have the option of reallocation of workloads to use as leverage. The competitive options provide the highest degree of leverage, because the allocation of future work is uncertain and depends on cost and technical performance.

Commonality of Ships. Using a single shipbuilder, either for the whole with the sole-source option or with portions of the ship with the block option, ensures that all 12 ships will have a high degree of commonality. With the directed-buy and competitive options, two different shipbuilders are producing ships and there may be differences, or a lack of commonality, across the ships in the Type 45 fleet.

Coordination and Integration of Multiple Shipbuilders. In addition to commonality of all the ships in the programme, the sole-source option results in the MOD and the PCO having to deal with only one shipbuilder. This aspect should result in improved coordination and integration during the total programme. With the other options, the MOD and PCO must interact with two shipbuilders, coordinating and integrating their activities to ensure that the programme stays on schedule and within budgets.

Coordination and integration become most difficult with the block options, for which the timing of the construction and transport of the blocks to the assembly site must be closely managed to ensure that there are no delays in the build schedule. Also, the dimensional control of the blocks must be closely monitored to ensure that the blocks *fit* correctly during assembly and that additional man-hours for rework are minimised or eliminated.

Collocation of Production and Support. Keeping VT involved in the Type 45 programme through either direct buys (of whole ships or blocks) or competition will mean that aspects of the production and support of the Type 45s will be collocated at Portsmouth. This should lead to reduced life-cycle support for the ship.

Risks and Rewards to Other Programmes

In addition to the Type 45 programme, the acquisition strategy chosen by MOD will affect, both positively and negatively, other shipbuilding programmes. As our analysis suggests, the sole-source option will result in reduced overhead costs for the Astute programme as more workload at BAE SYSTEMS Marine reduces overhead rates. Directed buys or competition will result in lower reduction of Astute overhead rates. The sole-source and block options, whereby the ships are assembled at Barrow-in-Furness, may also alter the Astute production process. Both the Astute and the Type 45 will be built in the Devonshire Dock Hall. With all 12 Type 45s going through the close confines of the DDH, chances will arise for scheduling problems or for minor accidents to disrupt either or both production lines.

If VT builds more whole ships through the directed buy of whole ships and the competition options, the potential for problems in the DDH would be lessened. Also, with the direct buy of whole ships and competition, VT remains a builder of warships. For future programmes such as the FSC and the CVF, VT's continued presence increases the chance of VT being in the market to compete for these programmes. Sole-source and, to some degree, the directed buy of blocks may not allow VT to maintain its warship construction capabilities, leading to future programmes facing a monopoly in warship construction.

Risks and Rewards to the UK Shipbuilding Industrial Base

Sole-source production of the Type 45s might have negatively influenced the UK shipbuilding industrial base. VT might have decided to exit the shipbuilding business or might have forgone its move to Portsmouth.

Directed buys and competition should help the UK shipbuilding industrial base become healthier and more robust. Although the future of BAE SYSTEMS Marine's Clyde shipyards could be threatened if it does not receive a significant portion of the Type 45 work, keeping VT active in building warships will be positive for future MOD programmes. A caveat must be stated for the option of a directed buy of blocks, however: Once that paradigm is chosen, it may be difficult to choose another paradigm for future programmes.

OVERALL INTEGRATION OF ANALYSIS AND OBSERVATIONS

The various advantages and disadvantages of the different acquisition options are summarised in Figure 4.8.

It is apparent from the figure that none of the options is dominantly superior to the others for Type 45 procurement. Sole-source procurement has an apparent advantage in near-term costs. However, those costs might grow in the absence of continuing competition,[6] and the option suffers in other areas. Selection from among the other options depends in large part on the weights the decision-maker assigns to a criterion relative to the weights for other risks or rewards in a column. Thus, this analysis is informative but not absolutely conclusive.

---

[6]See Appendix B for an estimate of why and how costs might grow if competition in naval shipbuilding is lost.

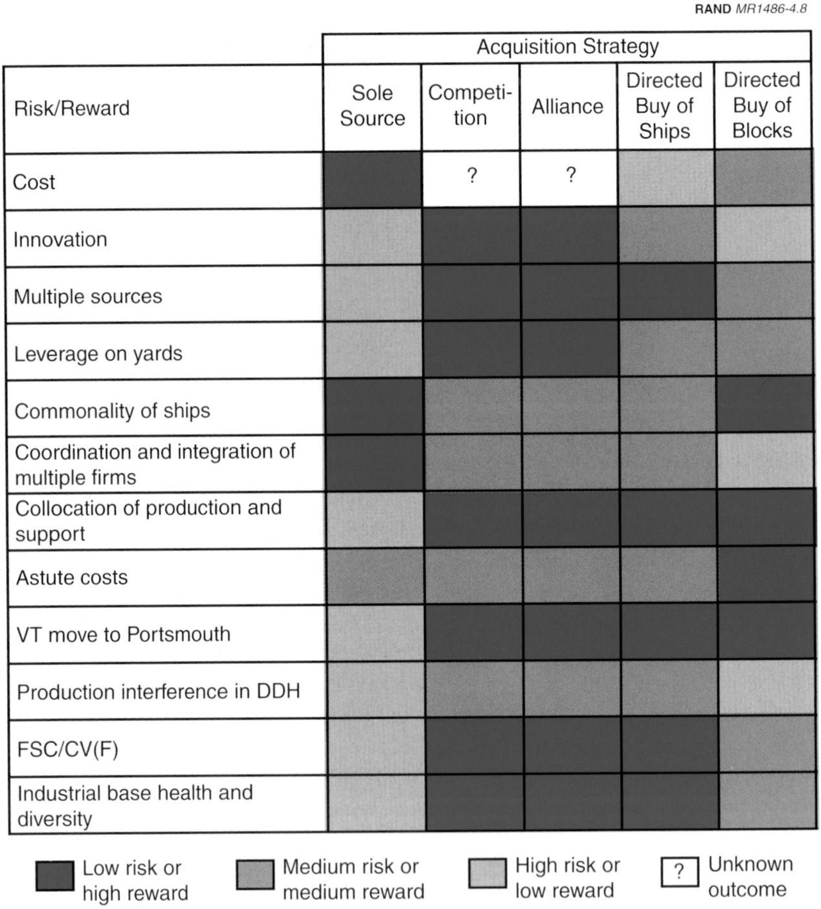

RAND MR1486-4.8

Risk/Reward	Acquisition Strategy				
	Sole Source	Competi-tion	Alliance	Directed Buy of Ships	Directed Buy of Blocks
Cost		?	?		
Innovation					
Multiple sources					
Leverage on yards					
Commonality of ships					
Coordination and integration of multiple firms					
Collocation of production and support					
Astute costs					
VT move to Portsmouth					
Production interference in DDH					
FSC/CV(F)					
Industrial base health and diversity					

Low risk or high reward   Medium risk or medium reward   High risk or low reward   ? Unknown outcome

Figure 4.8—Summary of Risks and Rewards of Alternative Acquisition Strategies for the Type 45

Chapter Five

THE APPROVED TYPE 45 PROGRAMME

Members of the RAND research staff presented the analysis described in the preceding chapters to senior managers in the MOD during the first half of June 2001. On 10 July, the Secretary of State for Defence, Mr. Geoffrey Hoon, announced the government decision on the programme to the House of Commons:

> . . . Working with the companies, we have developed a revised strategy, which allocates work on the ships between the two shipbuilders for the whole class of type 45 destroyers. The first-of-class ship will be assembled and launched at Scotstoun. The focus of design support to the whole class will remain there, with continuing participation by both shipbuilders. The remaining ships will be assembled and launched at Barrow-in-Furness.

> Vosper Thornycroft at Portsmouth, and BAE Systems Marine—on the Clyde and at Barrow-in-Furness—will both build and outfit substantial sections of each ship. The yards will continue to build the same sections throughout the programme, to increase efficiency and produce better value for money for the taxpayer.

> Under the strategy, a commitment has now been made to six ships in a planned class of up to 12 ships. That commitment has therefore doubled the number on order. This larger volume of guaranteed work, and a stable foundation to the project, will allow industry to make long-term investment decisions.

> Subject to negotiations being completed satisfactorily, I propose to adopt this revised approach, through which we are confident that

we can secure demonstrable value for money. We are seeking demanding efficiency improvements from industry. The initial findings of the RAND study support this new approach. It reflects the best features of the BAE Systems Marine bid in terms of learning from experience from one ship to the next, but it also preserves the possibility of competition for a number of subsequent defence programmes. The new strategy gives a welcome level of stability to our warship building industry. Above all, it offers the best prospect of achieving the in-service date for the type 45 destroyer, with deliveries to the Royal Navy starting in 2007. Any significant delay in that date would have significant operational and cost penalties.

The planned distribution of work between the shipyards is shown in Figure 5.1. The ship will be made up of six blocks plus the superstructure. Vosper Thornycroft will build two blocks (E and F) plus the superstructure; the BAE SYSTEMS Marine Clyde shipyards will build two blocks (A and D); the BAE SYSTEMS Marine Barrow shipyard will build two blocks (B and C). VT and the Clyde shipyards will transport their blocks to the Barrow shipyard, where the whole ship will be assembled (the first-of-class will be assembled at Clyde [Scotstoun]). BAE SYSTEMS Marine Barrow will conduct the ship trials, and Vosper will conduct the combat system trials. Table 5.1 details this allocation of work.

This solution offers a number of economic advantages. Spreading the Type 45 work between BAE SYSTEMS Marine and VT helps

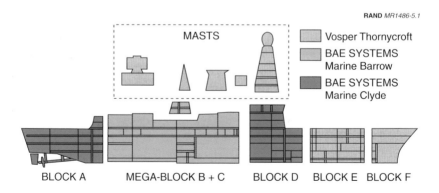

Figure 5.1—MOD's Type 45 Block-Production Allocation

Table 5.1

MOD Type 45 Work-Allocation Strategy Announced 10 July 2001

BAE SYSTEMS Marine Barrow	BAE SYSTEMS Marine Clyde	Vosper Thornycroft
First-of-class		
Perform design work	Design centre	Perform design work
	Produce all basic steelwork for Marine	Build forward section of the ship and masts/funnels, outfitted to about 80%, and barge to Clyde (Scotstoun)
	Build "Clyde" blocks outfitted to about 40%	
	Ship all "Clyde" blocks to Clyde (Scotstoun)	Conduct Stage 2 trials (combat system performance)
	Build "Barrow" blocks (main machinery spaces) from steel units from Clyde (Govan)	
	Receive VT elements; assemble and launch ship	
	Conduct Stage 1 trials (ship performance)	
Rest of class		
Build "Barrow" blocks (main machinery spaces) from steel units from Clyde (Govan)	Design centre	Perform design work
	Produce all basic steelwork for Marine	Build forward section of the ship and masts/funnels, outfitted to about 86%, and barge to Barrow
Receive "Clyde" blocks from Clyde (Govan) and forward section and masts/funnels from VT; assemble and launch ships	Build "Clyde" blocks outfitted to about 80%	
	Ship all "Clyde" blocks to Barrow	Conduct Stage 2 trials (combat system performance)
Conduct Stage 1 trials (ship performance)		

ensure that both shipbuilders will remain viable and able to compete on future MOD programmes. The strategy also facilitates the VT move to Portsmouth while helping secure the future of the Clyde shipyards. Having a single shipbuilder construct all of the same blocks for the Type 45 class takes maximum advantage of the learning effect and, therefore, reduces costs from what they would be if the construction of complete ships were distributed between the two shipbuilders.

Building blocks at multiple sites does have potential disadvantages. Additional costs will be associated with constructing the blocks with enough rigidity and weatherproofing to permit movements and transportation in such a way that the structural tolerances and integrity of the blocks will not be impaired. Structural tolerances must be managed very closely; misalignment of adjacent blocks can lead to substantial rework costs. Also, costs will be associated with transporting the blocks. Finally, scheduling of the construction and delivery of the blocks must be closely managed. A block that arrives late at the assembly yard may cause significant delays in not only the Type 45 programme but also the Astute programme.

This MOD decision will have direct consequences for shipyard modernisation actions. BAE SYSTEMS Marine will need to make investments at each of its facilities to accommodate the distribution of work shown in the table: The first Type 45, HMS *Daring*, will be assembled and launched at Clyde (Scotstoun). All subsequent Type 45s will be assembled and launched at BAE SYSTEMS Marine Barrow. The company's Clyde (Govan) shipyard will produce major steelwork for all of the ships.

In an interview shortly after the government's announcement, BAE SYSTEMS Marine's managing director, Simon Kirby, spelled out these investments ("Industry Update", September 2001, p. 42):

- **Barrow**—upgraded facilities for construction of Blocks B and C

- **Clyde**—fabrication facilities and block-transfer facilities (Govan); new door and cranes in the module hall, upgrades to the pipe shop, and new bending machinery (Scotstoun).

The revised Type 45 strategy will allow VT to proceed with plans to invest in a new shipbuilding facility within Portsmouth Royal Naval Base. Since spring 2000, the company had been planning to shift shipbuilding operations to Portsmouth, but had delayed committing to this move until the MOD's determination of its role in the Type 45 programme. Shortly after its role as a producer of major modules of the warship was clarified, VT announced that it would go ahead with the new facility, which will be built on four existing docks in the Portsmouth base. This construction will have the added benefit of reducing overhead costs on ship repair work to be performed at Portsmouth.

Chapter Six

ISSUES REQUIRING FURTHER STUDY

The Type 45 programme is one part of a larger acquisition programme laid out by the MOD for the Royal Navy and expected to extend over the next two decades. The decisions on the Type 45 procurement strategy will affect those future programmes, and the MOD will face a number of additional issues as the larger programme evolves. In this final chapter, we outline the following key issues remaining for the MOD to address over the coming years:

- What is the future manpower demand and supply picture?

- Is the distribution of ownership and management responsibility among the shipyards, PCO, and MOD appropriate?

- Which acquisition strategies are most viable for future programmes?

- How can innovation be encouraged in the absence of competition?

- How can core industrial base capability best be sustained?

- What are the issues when block production is performed separately from final assembly?

We explore these issues in turn in the following sections.

WHAT IS THE FUTURE MANPOWER DEMAND AND SUPPLY PICTURE?

The current and potential UK military shipbuilders with whom we spoke expressed confidence that they could rapidly expand their workforces to accommodate increased workload. They also said that this expanded workforce would require minimal, if any, training to become as proficient as current shipyard workers. We are less sanguine than they. Although there is a surplus of qualified shipyard workers today, the ambitious ship construction programme envisioned by MOD may very well exhaust the pool of experienced workers.

As discussed in Chapter Two, the MOD is embarking on an ambitious shipbuilding schedule in the coming years. After 2006, most of the military shipbuilding will be concentrated in just four programmes: Type 45, CVF, Astute, and FSC (see Figure 2.2). This concentration contrasts with the many smaller ship programmes being undertaken at present and will have profound implications for UK military shipbuilding. The smaller shipyards, which are incapable of producing these larger ships, will find themselves without a direct base of MOD work. To get smaller portions of work, such as modules, they will need to subcontract with the PCOs (not the MOD). How this subcontracting will play out is unclear at present, because one of the major PCOs has direct ties to several shipyards. By contrast, the larger shipyards should see strong workloads for the next decade, particularly BAE SYSTEMS Marine Barrow and VT.

Figure 6.1 depicts the demand for direct-worker labour, engineering, and support in military shipbuilding. The figure shows that the demand in that industry segment for direct workers will be fairly steady for engineering and support. Note, however, that because some shipyards include these workers as indirect labour, our values likely undercount real demand. Yet, the trend is clear. Demand for blue-collar workers doing direct labour declines slightly between now and 2006, after which demand increases rapidly to a peak of nearly 7,000 direct workers in 2010—almost double the level in 2005. Clearly, the main risk for the MOD's future programmes is maintaining and expanding the labour force in the next several years while managing a small dip in demand.

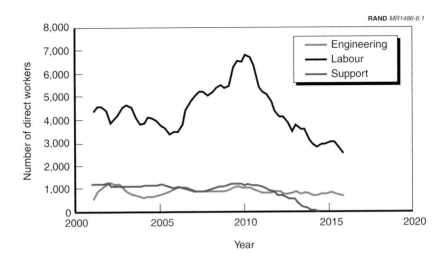

**Figure 6.1—Estimated Future Demand for Direct Shipyard Workers
to Produce UK Military Ships**

Data collected show that, as of fall 2001, direct-labour employment
stood at a little more than 5,000 workers. Given the projection from
Figure 6.1 that shipyards will need nearly 7,000 workers in 2010, there
will be a shortfall of nearly 1,800 workers if the shipyards do not increase staffing from current levels. Further complication the matter
is whether workers will be available in the geographic locations
where they will be needed. The shipyards were confident of being
able to attract workers from all over the country. We are less confident. Certain shipyards may find it difficult to staff appropriately.

Moreover, it is unclear how many workers will be available to move
to military shipbuilding from commercial shipyard work. In recent
years, commercial shipbuilding capacity in the United Kingdom has
fluctuated, with each succeeding peak lower by about 200,000 gross
registry tons (GRTs). Figure 6.2 shows commercial ship tonnage
produced by UK shipyards over the past two decades.

This figure does not necessarily reflect all UK shipbuilding. For example, it does not include warship building. But it illustrates the
problem: The tonnage being produced is declining. Estimates vary

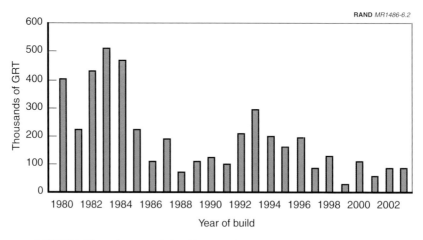

SOURCE: The values are derived from the Lloyd's Maritime Information System database of all *Lloyd's Register* ships built in the UK since 1980. Data are current through October 2000.

Figure 6.2—UK Commercial Shipbuilding Gross Registry Tonnage Versus Year

for the number of workers remaining in shipbuilding trades at this time. Armstrong Associates estimated recently that there are only about 4,000 workers left in the nonmilitary shipbuilding trades, and that there are another 10,000 working in building Royal Navy vessels (Thales, 2001). The latter value corresponds to our shipyard survey total of over 9,300. The UK Shipbuilders and Shipbreakers Association (SSA, 2001) is more optimistic, estimating that there are about 26,000 workers engaged in shipbuilding and ship repair.

Given such workforce issues, we recommend that the MOD consider the following four strategies:

1. Encouraging PCOs to include options whereby the smaller shipyards can compete for subcontracts through the PCOs or from the larger shipyards.

2. Requesting companies to keep industry informed on future plans so that the government can implement shipyard training programmes that will ensure adequate numbers of qualified

workers for the planned work; and assessing the effect of new workers on overall productivity and projecting that effect onto the cost and schedule of future programmes.

3. Detailing issues regarding the availability and cost of labour as a function of the construction site selected and specifying how regional difficulties, if any, are to be overcome. Labour shortages could drive up labour rates substantially.

4. Changing the start dates of future programmes to shift or lessen the peak demand for workers.

IS THE DISTRIBUTION OF OWNERSHIP AND MANAGEMENT RESPONSIBILITY AMONG THE SHIPYARDS, PCO, AND MOD APPROPRIATE?

Both the Type 45 and Astute PCOs are part of BAE SYSTEMS, as is one of the PCOs competing for the CVF. The PCO concept envisions an organisation independent of shipyards among which the PCO is expected to conduct a vigourous competition. BAE SYSTEMS also controls three shipyards that perform the majority of military ship construction. Many of the non–BAE SYSTEMS personnel and shipyard management teams with whom we spoke were quite critical of having the PCOs housed in the same company as the shipyards, and clearly did *not* have confidence that sufficient internal company firewalls could be erected and/or sustained between the BAE SYSTEMS' PCOs and shipyards.

The appearance of a conflict of interest and possible compromise to fairness, if not an actual conflict of interest, raises an issue that needs additional consideration and vigilance by the MOD.

In addition, the PCO approach is a new acquisition strategy for the MOD. As experience is gained, further modifications/improvements in this strategy may be required.

WHICH ACQUISITION STRATEGIES ARE MOST VIABLE FOR FUTURE PROGRAMMES?

After examining the relative costs and other consequences of four different acquisition strategies applied to the Type 45 procurement—

sole-source, directed buy of whole ships across two firms, competitive procurement of whole ships, and directed buy of ship blocks—we drew two conclusions in view of the characteristics of the Type 45 programme and the present status of the naval shipbuilding industry in the UK: No single strategy was predominantly superior, and selection of a preferred strategy depends on the judgement of senior managers in assigning relative importance to various strategy attributes. However, these conclusions do not necessarily apply to other programmes to be conducted in the future. No particular procurement strategy can be claimed as best until the particular situation has been examined.

The selection of a preferred strategy for future programmes will depend, in large part, on two factors: number of ships to be procured and distribution of other business across the available shipyards.

Number of Ships to Be Procured

Because of loss of learning effects, splitting large production runs among two or more producers can incur large cost penalties that might, or might not, be recovered through competitive pressures. For very small production quantities, say, two or three ships, the absolute cost of learning effects is smaller; other shipyard costs, such as start-up investment and sharing of overhead, become more important.

Distribution of Other Business Across Available Shipyards

Ship construction is labour-intensive, but labour is not readily transferred from one shipyard to another. Thus, the costs and benefits of placing a particular ship-construction project at a particular shipyard can depend heavily on the shipyard's other current and projected business.

HOW CAN INNOVATION BE ENCOURAGED IN THE ABSENCE OF COMPETITION?

Consolidation within the military shipyards, which reduced overhead and facility costs, has resulted in significant savings for MOD. However, a more concentrated defence sector may be less innovative, the smaller number of remaining firms perceiving a reduced need for new ideas in order to win contracts. Contractors may also be reluctant to pursue innovations that may "cannibalise" lucrative existing markets (i.e., they do not want to develop competing products within the existing market). Increased MOD emphasis on cost control may discourage firms from undertaking technically risky options out of fear of possible failures and consequent adverse costs. Also, with less emphasis on innovation, contractors may assign their best staff and most fertile minds to tasks other than advanced defence R&D; the best minds, in turn, may seek out smaller, more entrepreneurial firms in other industries, thereby leaving the defence industry altogether.

Arguably, reduced acquisition budgets make innovation more important than ever. Smaller military forces must perform more demanding roles. Hence, MOD must be, and is, interested in strategies for encouraging innovation in this changing defence sector.

RAND studies of military contractors' behaviour (Birkler et al., 1994, 1998) suggest that, in the past, innovation has come from two sources: (1) a large number of independent contractors engaged in dynamic and intense competition to win the major contracts and (2) a second tier or marginal set of prime contractors willing to take more technological and financial risks in order to break into the first tier of prime contractors. As shipbuilding has consolidated, these traditional sources of innovation have disappeared or are in the process of disappearing. Even more worrisome, the new consolidated primes usually reduce R&D spending and the number of suppliers even more.

Many industrial sectors in the United States have seen significant consolidation in recent years, but without a concomitant reduction in innovation. Production (and marketing) activities are increasingly consolidated in the pharmaceutical and biotechnology industries, in the telecommunications sector, and in some segments of the computer hardware and software industries. In all of these sectors, technological innovation has nonetheless continued at a very rapid pace. Popular wisdom suggests that much of this innovation can be attributed to the separation of production from R&D in each of these sectors for practicality. Small, R&D-oriented firms are easily established and compete fiercely to produce the next technological advance. Successful innovators license or sell their discoveries or are themselves bought by the few consolidated firms that dominate production and marketing.

A version of this pattern of innovation has been seen in the cruise-ship industry, in which subsystem contractors are competed, with the effect that technical risks and costs to the primes are reduced.

The MOD has an interest in capturing the efficiencies that come from consolidation of the military shipbuilding production base *and* from sustaining innovation. The MOD would be well served by gaining a better understanding of the effects of defence industry consolidation on innovation and of how other rapidly consolidating industries have managed to sustain and encourage robust innovation.

HOW CAN CORE INDUSTRIAL BASE CAPABILITY BEST BE SUSTAINED?

Although investments are still required, the Type 45 decision sustains the industrial base over the next decade for its size and class of warship. BAE SYSTEMS Marine is in the process of making capital investments and reorganising and optimising facilities for its current workload. By moving to Portsmouth, VT can construct larger ships, on the scale of the Type 45, and blocks than were possible at its Woolston facility and keeps open the opportunity to participate in Type 45 production and to compete for other programmes in the future. Thus, the Type 45 solutions enhance and preserve vital elements of the industrial base. However, the availability of a qualified workforce as the programme moves to rate production is still an active issue, especially in light of the workforce required for the CVF programme.

The proposed size of the future carriers is believed to be beyond the current production capability of any single UK shipyard. The current plan is that major portions, or blocks, of the carriers will be constructed in several shipyards and transported to one shipyard for final assembly. But only a few facilities are large enough to assemble the ships, and each has shortfalls and constraints.

In addition to the problem of facilities and capacities within the current shipbuilding industrial base, any construction plans for future carriers must include an assessment of how other current and future shipbuilding programmes, both military and commercial, will curtail or enhance the availability of workers and facilities.

Finally, given the potential need to involve and integrate several shipyards in carrier construction, the MOD must resolve several major issues: the scheduled start and completion of the carriers, the gap between the start of the first and second ships, and the need for an ensured facility for subsequent life-cycle support.

WHAT ARE THE ISSUES WHEN BLOCK PRODUCTION IS PERFORMED SEPARATELY FROM FINAL ASSEMBLY?

Shipyards worldwide routinely build, outfit, and join together modules of 1,000 metric tons every day. Many very large marine structures have been built in ultra-large modules, which are subsequently assembled to form one structure. But all of the examples of which we are aware are far simpler structures than a surface combatant or aircraft carrier such as CVF. For both the Type 45 and CVF, subassembly fabrication and block construction will be done at locations remote from the final assembly point. In our judgement, the larger and more complex the modules to be joined, the greater will be the difficulty of ensuring adequate mating and the risk of costly rework. The shipyards must provide enough planning and analysis detail to support confidence on three issues connected with modular construction: block fabrication and assembly, out-of-sequence delivery of blocks, and time on assembly berth. Table 6.1 outlines these issues and the actions that the MOD should require of the PCO.

Table 6.1

MOD-Required PCO Actions for Type 45 Block Construction

Issue	MOD-Required Actions
Block fabrication and assembly	Demonstrate that the detailed design process, construction process, and attendant transportation, unloading, and mating processes for each block ensure that the blocks will fit together; additional rework will be minimal; and that successful delivery of the ships will result within estimated costs and budgets.
Out-of-sequence delivery of blocks[a]	Describe the delivery sequence of blocks intended for the PCO's selected production scheme and how delivery delays of any component will be accommodated. This description shall include how out-of-sequence delivery will be accommodated at the final assembly yard and how the accommodations affect cost and final ship delivery.
Time on assembly berth	Show, by detailed analysis of the work required, how the assembly in-dock will be accomplished in the time allowed in the dock.

[a]Understanding out-of-sequence delivery of blocks is especially important for the Type 45 programme, since the Type 45 will be assembled in the DDH, along with the Astute-class submarines.

SENSITIVITY ANALYSIS

Throughout our evaluation on the acquisition options for the Type 45, we made many assumptions about the UK industrial base (as described in the main text). Without these assumptions, it would not have been possible to make quantitative evaluations. However, the reader might question the degree to which our results would change if different assumptions were made. How robust are the results? Do small changes in the assumptions dramatically change the conclusions? To determine how robust the results are, we change a number of key assumptions about number of ships acquired, learning slopes, CVF workload, and workforce productivity to see whether the outcomes are dramatically different.

NUMBER OF SHIPS ACQUIRED

We have assumed throughout the main body of the analysis that 12 Type 45 ships will be built over the next two decades. However, the total number is far from certain. It will depend on many factors. Typical of military acquisitions, budget shortfalls result in fewer systems being purchased than originally planned. Likewise, priorities may change over the long production time, leading to fewer or more systems being required. To understand the effect of changing the number of ships acquired during the Type 45 programme, we varied the number from three to 15 total ships.

Figure A.1 shows the cost penalty (over the entire Type 45 programme) to break even for an acquisition programme of nine ships. The values are somewhat lower than those observed for the 12-ship programme (see Figure 4.7).

Figure A.2 shows the cost penalty for the Alliance strategy at various build levels for the Type 45 only and for all programmes. As can be seen in the figure, the cost penalty increases with the number of ships, but nonlinearly. It tends to flatten as the number of ships increases.

This does not mean that it is easier to achieve savings through competition as the number of ships in the programme decreases. In fact, the reverse is true for the Alliance strategy. Recall that, under the Alliance strategy, the first three ships are allocated and the remaining ships are competed in lots of three. Thus, the benefits of competition are not fully gained until the second and subsequent lots. If we adjust the cost-penalty levels to account for the number of ships actually competed, the trend is rather flat but *decreases* somewhat with the quantity (Figure A.3; see also Appendix B).

LEARNING SLOPES

For the production of the Type 45, we assumed the learning slopes to be those that BAE SYSTEMS Marine and Vosper Thornycroft re-

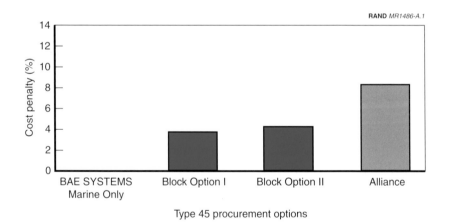

Figure A.1—Cost Penalty for a Production Run of Nine Ships

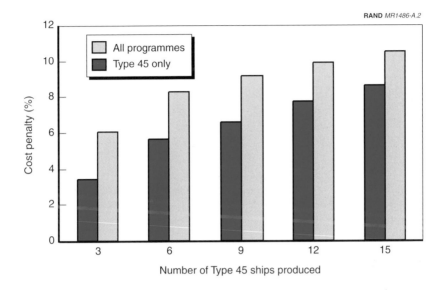

Figure A.2—Cost Penalty for the Alliance Strategy for Varying Levels of
Procurement

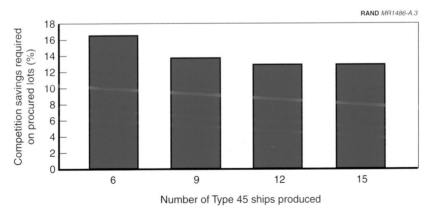

NOTE: We included a production run of 15 Type 45s to better understand the cost
sensitivity connected with a large total programme size.

Figure A.3—Competition Cost Penalty for the Alliance Strategy for Varying
Levels of Procurement

ported, which were not identical. The difference was not surprising, considering that the two yards have different work-breakdown and cost structures. Would our results change if we assumed an equivalent learning slope at each yard?

Figure A.4 shows the cost penalty for each of the seven strategies and different learning assumptions. For the common learning slope, we assumed a 90-percent slope, which is slightly higher than the average of the values from the two yards. For this common case, the cost penalty is lower by approximately 1 to 4 percentage points, depending on the strategy. All values are within 8 percentage points. This lowering of the cost penalty is consistent with the overall higher learning slope discussed in Chapter Four.

CVF WORKLOAD

Another assumption we made was that VT would perform one-quarter of the CVF production work. In the data provided to RAND, VT did not speculate on the level of work for CVF production,

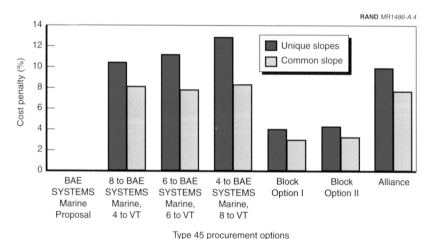

Figure A.4—Cost Penalty with Different Learning Assumptions

because the programme was in the early design phases. Do the results change if we remove the CVF workload from VT? According to Figure A.5, the answer is, "Very little".

WORKFORCE PRODUCTIVITY

In other work by RAND (Birkler et al., 1994, 1998), the authors have found that productivity changes in the workforce can dramatically alter the cost of production for ships. For the study on the Type 45, we assumed that new, unskilled workers start at 67 percent proficiency and linearly improve over time to 100 percent at the end of three years. Furthermore, we assumed that the pool of new hires has an equal distribution of workers over the proficiency range of 67 to 100 percent.

However, the shipbuilders opined that getting additional, fully productive workers would not be difficult. In essence, they anticipate that all new hires will be fully productive. Figure A.6 shows that these assumptions on productivity do not change our results substantially.

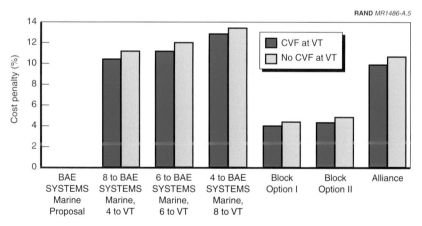

Figure A.5—Cost Penalty with Different CVF Workload Assumptions at VT

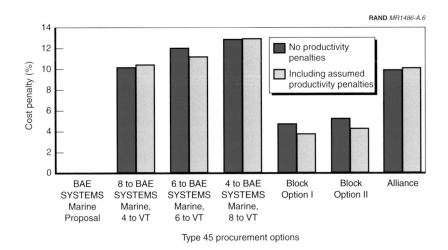

Figure A.6—Cost Penalty with Different Workforce-Productivity Assumptions

LONG-TERM IMPLICATIONS OF A LACK OF COMPETITION

Our analysis of the cost and benefits of the various acquisition strategies for the Type 45 programme looked only at the short-term implications. For the BAE SYSTEMS Marine unsolicited offer (sole-source), the production cost was lower than that for the other directed-allocation strategies. However, although the sole-source might appear to be the least expensive procurement strategy in the short term, it poses considerable long-term drawbacks. The primary drawback is that selecting the sole-source strategy could result in a lack of competition in UK surface combatant production (BAE SYSTEMS Marine will have a virtual monopoly). How might a lack of competition affect ship procurement costs in the long term?

In Chapter Four, we observed that competitive ship and missile programmes, on average, cost about 7 percent less than noncompetitive programmes (see Figure 4.3).

Another cost of noncompetitive industries is a higher rate of price escalation. Figure B.1 shows the consumer price growth in five different industries: drugs and pharmaceuticals, cars, electricity, petroleum (gasoline), and airfare. As a reference, we also show the Consumer Price Index (CPI). We obtained industry and CPI data from the U.S. Bureau of Labor Statistics. As is readily evident, the drug and pharmaceuticals, and airfare indices have the highest rate of escalation over the 11-year period. They averaged 1.7-percent greater annual escalation than the CPI. We use this 1.7-percent additional escalation rate as an example to illustrate the possible consequences of reduced competition in future UK shipbuilding.

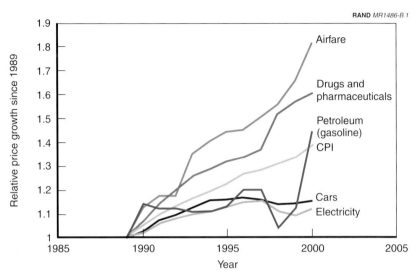

SOURCE: U.S. Bureau of Labor Statistics.

Figure B.1—Relative Price Growth in Various Industries from 1989 to 2000

To quantify the effects of greater procurement cost and greater price escalation on the MOD, we assumed that warship production after completion of the Type 45 programme (about 2014) would continue at the annual investment rate of that year and would be noncompetitive. All procurement costs from that point would be 7 percent higher and escalate at a 1.7-percent higher rate. When doing a cost-benefit analysis of potential future savings, an accepted practice is to use techniques such as discounting (net present value) as a metric to quantify benefits. This method balances the fact that cash flows do not occur in the same year. Because there is an opportunity cost of capital, receiving £1 today is more desirable than receiving £1 five years from now. This desirability does not relate to inflation; rather, it reflects the fact that if we forgo use of capital, we lose the ability to spend it on something else. Likewise, we gain benefit if we have the ability to use it sooner. Therefore, it is an *opportunity cost.*

Figure B.2 shows how the net cost savings from sole-source production erodes relative to a strategy whereby each shipyard builds six ships, thereby maintaining two shipyards that could compete for future programmes. The net production savings for the sole-source case is the cumulative savings through a sole-source procurement (negative means savings). It is generally cheaper to produce in the sole-source case, but that savings gets eroded through the greater escalation and the future greater procurement costs (the 1.7 percent and 7 percent, respectively). *Undiscounted* means that no discount rate was applied (0 percent). The 4-percent discount rate means that a cost of capital (discount rate) of 4 percent per annum was applied.

Two curves are shown. The black curve displays the net savings undiscounted; the blue curve shows the net savings discounted at 4 percent per year.

Notice that the most savings is achieved by 2012–2013 (the end of the Type 45 production). After that point, ship-production cost becomes more expensive (owing to a lack of competition). The break-even points, at which all the savings from sole-source production is eroded, are about 2021 undiscounted and 2026 discounted. From then on, the sole-source strategy for the Type 45 results in higher costs for future programmes.

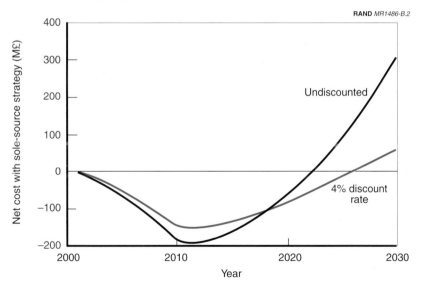

Figure B.2—Net Cost Resulting from a Lack of Competition

BUILDING SHIPS IN BLOCKS AT MULTIPLE SITES: IMPLICATIONS FOR THE TYPE 45 PROGRAMME[1]

One of our research tasks was to better understand the extent to which a strategy of building ships in large blocks at multiple sites is used in commercial and military shipbuilding. Another task was to describe the potential advantages and disadvantages of block construction for the Type 45 programme. This appendix presents the results of our survey of current shipbuilding practices, describing precedents for block and multiple-site construction.

This appendix uses the following terminology:

- *Interim product*: a level of the product structure; it is the output of a work stage and is complete in and of itself.

- *Part*: the lowest level of the interim product structure. It can be either manufactured by the shipyard or purchased. Examples of shipyard-manufactured parts are a part cut from a plate, a stiffener cut from a tee profile, and a pipe piece cut from a pipe length. Examples of purchased parts are main engine, steering gear, and pumps.

- *Subassembly*: a small interim product, relative to an assembly, that is made from parts. It generally consists of one or a few plate parts with stiffener flanges connected to them.

---

[1]This appendix summarises material prepared for the study by Tom Lamb, P.E., EUR ING, Technical Associate, Innovative Marine Product Development, LLC, Ann Arbor, Michigan.

- *Assembly/panel*: an interim product that is made from sub-assemblies and parts. It generally consists of a single skin (plating) that has stiffeners and web frames connected to it.

- *Unit*: a structural interim product made from assemblies, sub-assemblies, and parts. It generally is a three-dimensional structure, such as the shell, bulkheads, and decks, having assemblies joined perpendicularly to each other.

- *Block*: a number of units joined together.

- *Grand block*: a large ship section made from two or more blocks before it is erected in the building berth.

- *Ring section*: a block or grand block that extends from the keel to the main deck and from one side of the ship to the other.

- *Machinery module*: a group of outfit items made into a self-contained package consisting of the support framework, grating, equipment, pipe, controls, etc.

- *Zone*: a geographical portion of the ship, such as the bow or a package of machinery.

In the next section, we employ these terms to describe the advantages and disadvantages of block construction.

BLOCK SHIP-CONSTRUCTION STRATEGIES

Until about 1950, ships were generally built piece by piece (similar to a building site), an approach with a very low investment cost, minimum crane requirements, and small-scale transport. Today, the piece-by-piece approach is rarely used, except in undeveloped countries, where the labour cost is low, and in developed countries for some small craft and one-of-a-kind ships. Figure C.1 shows the piece-by-piece construction of a small craft.

Over the past half century, shipbuilders have recognised the advantages of building larger portions of ships in covered production facilities, then assembling those portions in a dry dock or on a slipway. As larger numbers of similar ships are built, standardisation, repetition, and automation lead to economies of scale and production ef-

RAND *MR1486-C.1*

Figure C.1—Piece-by-Piece Construction of a Small Craft

ficiencies, and, in turn, to lower costs and reduced schedules. But, as the size of the modules built in production facilities grows, more investment is needed for larger cranes and transporters, and much more attention is required for configuration control to maintain the build tolerances between the modules.

Assembly Construction

The ship-construction method that erects assemblies on building berths, shown in Figure C.2, is common for small- to medium-sized ships. It has the advantages of being somewhat faster and less expensive than piece-by-piece construction, requires minimal investment, and provides greater flexibility in the sequencing of construction. It has the disadvantages of still being fairly slow, having low productivity and, hence, relatively high labour costs, and having the need for rework built in.

RAND *MR1486-C.2*

Figure C.2—Assemblies Being Erected on Berth

Two sites where assembly ship construction takes place are

- Appledore Shipbuilders in Devon: one of the first modern covered shipyards, it uses the assembly build technology typical of 1970 (when it was developed).

- ASEA shipyard in Sestao, Bilbao, Spain: an old riverbank shipyard, it has been modernised and extended for building larger, modern ships. It is limited by the need to construct from relatively small assemblies.

Block Construction

Block construction, common for large vessels, also is used for some smaller ones. Blocks can vary in size from approximately 50 tons for small vessels to up to 400 tons for large vessels such as very large crude carriers (VLCCs). A typical block is shown in Figure C.3.

Block construction has advantages that are more dramatic than those for assemblies: higher productivity and, therefore, lower labour costs. But blocks also entail disadvantages: the need for highly accurate assembly, a larger investment cost in facilities, and a very high reliance on control of accuracy and on on-time delivery of materials.

Frigates being designed and constructed in Spain for the Spanish and Norwegian navies use the block-construction approach, as shown in Figure C.4. However, these ships are employing a more traditional block approach, in which many small blocks (up to 100 tons) are fabricated in the same shipyard.

RAND *MR1486-C.3*

Figure C.3—Typical Structural Block

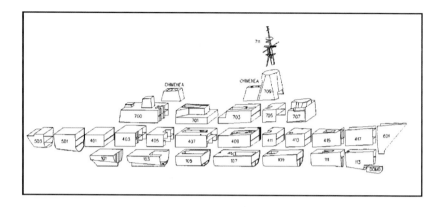

Figure C.4—Spanish Frigate Structural-Block Breakdown

Other examples of block construction include the following:

- Kvaerner Masa, Helsinki: a very effective builder of passenger ships that uses substantial blocks (with significant pre-outfitting) to exploit its covered dock.

- Astilleros Espagnoles, Cadiz: a classic large, greenfield shipyard[2] originally developed for VLCCs. It is equipped with Goliath cranes capable of lifting 600 metric tons each.

Grand-Block Construction

Grand blocks can be built from either assemblies or smaller blocks. Grand blocks are usually outfitted and painted in advance (see Figure C.5). The benefits attributed to grand blocks are reduced building-berth erection time and welding, easier access to blocks being assembled into the grand blocks, and no need for staging on the building berth.

One disadvantage of grand blocks is that they need to be moved to the building berth via large-lift-capacity cranes or other means. They also need to be aligned to other grand blocks.

---

[2]A *greenfield shipyard* is a shipyard constructed from scratch on a site that did not previously have a shipyard.

RAND *MR1486-C.5*

Figure C.5—A Typical Grand Block, Comprising Seven Blocks

Ring Construction

Generally used with mid-sized vessels, the ring approach is not as common as block construction. It has been used for some large ships as well, as shown in Figure C.6.

The several advantages of ring construction are that it can be substantially completed under cover; it improves productivity and lowers labour costs; and it allows production from assemblies, which provide flexibility to recover from inaccuracies, or from blocks. The disadvantages of ring construction are that it requires substantial investment costs and has an inflexible sequence of work, making it totally dependent on timely deliveries.

RAND MR1486-C.6

Figure C.6—A Typical Ring Section

Sites employing ring construction include the following:

- Electric Boat and Newport News Shipbuilding in the United States, and Barrow in the United Kingdom: both build submarines by the ring approach.

- Litton Industries, Pascagoula, Mississippi: developed in the 1960s specifically to exploit this method of building for frigates and destroyers.

- BAE SYSTEMS Marine Clyde (Govan): built liquid-nitrogen gas (LNG) ships from ring sections of up to 3,000 metric tons. BAE SYSTEMS Marine Barrow has since adopted ring construction for the Fleet Replenishment Ship (AOR) and Landing Platform Dock (LPD) contracts for the UK MOD.

EXAMPLES OF MULTISITE CONSTRUCTION

Once the techniques of building ships in large blocks were mastered, several shipyards took the next step of building blocks at multiple shipyards, then transporting them to a single shipyard for assembly into a whole ship. This technique of assembling ships from large blocks produced at different locations is more common than might be expected, and it has both potential benefits and problems.

World War II Liberty ships may have been the first example of ships constructed from pieces built at multiple shipyards. To produce large numbers of ships in a short timescale, conventional shipbuilding techniques were not adequate. As the demand for naval ships burgeoned, existing shipyard capacity could not keep pace. The use of welding and prefabrication delivered the solution—assemblies and pre-outfitting. Liberty ship sections were pre-assembled and transported to an existing shipyard for final assembly and launching. Despite the huge publicity about the success of these methods, the costs were high and rework was common. However, these ships demonstrated the feasibility of the method and met demanding ship supply and delivery schedules.

After the end of World War II, shipbuilding returned to more "traditional" practices. Occasionally, programmes used multisite block-construction techniques. For example, in the early 1970s, Seatrain Shipbuilding started the construction of large tankers while the shipyard in the old Brooklyn Navy Yard was being refurbished. To maintain its desired production schedule, Seatrain had all the structural blocks for the cargo tanks for the first ship, and many for the second ship, constructed by US Steel in Orange, Texas, and barged to New York. Also at this time, Litton built the bow and stern of a Great Lakes bulk carrier, joined them together, and sailed the joined sections to Lake Erie, where they were cut apart and attached to the cargo mid-body, which was built in a Great Lakes shipyard.

Other examples of building sections of ships at multiple shipyards include the following:

- Swan Hunter used multisite production to manage a large programme of tanker production in the mid-1970s, during which time the company had several shipyards on the River Tyne downstream from Newcastle, each specialising in different sizes and types of ships. For the tanker programme, which had a large number of ships and a short timescale, Swan Hunter pooled its resources through multisite construction. It developed new facilities at the Hepburn shipyard to provide a major new steelwork facility built around an existing dry dock. It also used existing, smaller steel facilities at the nearby Walker shipyard. To meet the programme schedule, the tanker aft ends were constructed and outfitted at Walker and towed to Hepburn, to be joined to steel-intensive cargo sections and a bow that had been constructed there. The forebody was built in the dry dock, taking advantage of the new and extensive steelworking facilities. No major problems were reported in aligning or in completing the ships once the dock had been drained.

- In the 1980s, Mitsubishi Koyagi, Hitachi Ariake, and Ishikawa-jima-Harima Kagoshima built eight large floating crude-oil reservoir tanks. The IHI Kagoshima Works supplied all the large blocks for the project, and final assembly was done at Koyagi and Ariake, facilities located near each other in Kyushu. The blocks were barged from the Kagoshima manufacturing facility to the Ariake and Koyagi erection sites.

- In Spain, the Sestao shipyard within the old Astilleros Espagnoles group set up subcontractor facilities on its land, where it could build structural blocks, thereby enabling it to reduce the number of permanent employees.

- BAE SYSTEMS Marine (formerly Marconi-Marine) has used multisite construction to build conventional (SSK) and nuclear (SSBN) submarines. Specialised production facilities were available in two sites for the important pressure hull rings (of differing sizes and configurations). To avoid wasting time or capacity, BAE SYSTEMS Marine decided to manufacture the hull-ring units at the most convenient site and transport them 100 miles by road to the other facility, as required. The solution was successful, but probably at some additional cost, and the finished submarines were satisfactory. Most important, the peaks and troughs in production were avoided and the programmed schedule was maintained.

- "Jumboising" is fairly common in ship conversion. Typically, a vessel to be enlarged is dry-docked and cut in half transversely through all the structure and systems. Thereafter, the ship halves may be separated and moved via hydraulic or other ground-transfer systems to other parts of the dry dock. The new mid-body is built unit by unit, depending on crane capacity and the space between the ship halves filled in by the newly built assemblies.

- The Canadian frigate programme, which took place from the late-1980s through the mid-1990s, used some multisite work. The main contract was placed with St. John Shipbuilding in New Brunswick. However, the programme had to have a second source and St. John had to subcontract some of the ships to Marine Industries Limited (MIL), based in Quebec. MIL had two sites at Quebec City and Sorel-Tracy. To meet the tight-programmed requirements, MIL used both sites for blocks, but completed final construction at Davie. Transport was of substantial, floating hull sections, which moved approximately 150 miles on the St. Lawrence River.

- A programme around 1990 to construct new ferries for British Columbia required a capacity for ship construction that was not available locally in existing shipyards. To fulfill the strong desire to retain the programme locally, an imaginative solution was developed. Three sites were identified, and the ferry construction was divided into three ship sections. The hull was split into fore and aft parts, and the superstructure formed the third major section. Each section was constructed at a separate site, after which the hull halves were floated together and the superstructure was put in place. The process was achieved with a limited expenditure for a single, two-ship contract. In fact, simple technologies were used imaginatively to move substantial ship sections. However, the approach was not cost-effective: The original contract price was significantly more than the single-shipyard bids from other Canadian and U.S. shipyards, and the project's final cost was more than 216 percent of the contract price.

- Newport News Shipbuilding subcontracted all of the deckhouses for the Double Eagle product tankers it built in the 1990s. However, substantial rework costs were incurred when the deckhouses did not align properly with the ship body.

- To build the Disney cruise ships, Fincantieri selected two shipyards previously not involved in the cruise-ship business. The shipyard near Venice had both block-construction and final-assembly responsibilities.

- The Netherlands shipyards organise in geographical clusters of shipyards that have considerable internal competition for domestic orders. However, they have a high degree of collaboration for export orders. Such collaboration involves large-scale interyard subcontracting. As a result, relatively small shipyards are able to obtain or share in significant orders, especially in terms of the numbers of ships constructed. The shipyards are also able to offer rapid delivery dates. One shipyard may accept an order, then subcontract various blocks to other shipyards within the cluster. Such subcontracting enables each to specialise in a particular block type. The process is applied to relatively small, simple ships, but is extremely effective in minimising costs and reducing timescales.

- DANYARD in Denmark established a process whereby it constructed the mid-bodies of its ships in the Aarhus shipyard and towed them after launch to the Fredrikshaven shipyard, which had built the bows and sterns in parallel (Figure C.7). The deckhouses were subcontracted to a Polish shipyard (Figure C.8). The Fredrikshaven shipyard connected the parts and completed and delivered the ships.

- Imabari's new, large Saijo shipyard opened in March 1995 as a dedicated block-production centre with no final assembly capability. It was at that point "a new generation factory" built with the idea of realising increased productivity and accuracy by "complete introduction of most advanced automatic equipment and apparatus" (sic) (Cooperative Association of Japanese Shipbuilders, 1996). Saijo's building dock became operational in 2000, and the plant is now a full-fledged shipyard.

There are other examples of using multiple shipyards in the construction of ships. The following describe cases where one shipyard builds the basic ship (i.e., the Hull, Mechanical and Electrical [HM&E]) and a different shipyard performs many of the system-outfitting functions:

- The hull for the UK helicopter carrier, HMS *Ocean*, was constructed in BAE SYSTEMS Marine Clyde (Govan) shipyard and sailed to BAE SYSTEMS Marine Barrow for combat system outfit-

RAND *MR1486-C.7*

Figure C.7—Stern Grand Block at DANYARD in Denmark

ting. This arrangement was reportedly based on a cost decision, but was also necessitated because Clyde (Govan) had the better steel-production facilities for this type of ship.

- Many Floating Production, Storage and Offloading (FPSO) ships are built in Asia but are sailed to the United States or Europe to have all of the topside equipment installed. The Asian shipyards do not have the experience to install such complex topside equipment for the offshore oil-drilling industry.

- Mitsubishi in Japan constructs hulls for warships in a small, traditional shipyard with end launching ways. It then tows the hulls about 15 miles to the main Nagasaki shipyard, where they are completed. This process serves to keep Mitsubishi's warship building as separate as it can be from its other shipbuilding activities.

RAND MR1486-C.8

**Figure C.8—Subcontracted Deckhouse Constructed in Poland for
Denmark's DANYARD**

Now that we have seen how shipbuilders spread work around multiple shipyards, we look at why they do so.

RATIONALE FOR MULTISITE CONSTRUCTION

Shipbuilders choose to spread work among various shipyards for many reasons. Political motivations to spread work among different areas of the country or to maintain an adequate number of firms in the industrial base may be one reason. A number of recent U.S. Navy shipbuilding programmes, the DDG 51 and LPD 17 classes and the Virginia-class submarine programme, for example, have been split between two shipyards.

Demanding delivery schedules may also force shipyards to build various portions of ships at multiple sites simultaneously. In the shipbuilding industry, demand is highly variable; so, too, are the odds that a shipyard will win an order. As a result, shipyards are reluctant to maintain workforces sized to meet the highest demand.

The flexibility to obtain blocks from other fabricators in such market conditions offers shipbuilders a definite advantage.[3]

The most compelling reason for shipbuilders to build portions of ships at multiple shipyards is the potential for reduced costs. Theoretically, it is possible to generate cost savings by concentrating specific blocks in one company so that duplication of skills and facilities can be eliminated. The result is lower overall overhead costs, as well as the benefits of learning extended over a longer production run. Some recent decisions on the subcontracting of structural blocks have been driven by this cost-savings focus. Some very efficient new facilities focused on building only blocks for other shipyards to assemble have been introduced in Japan and Europe.[4] The basic motivation for this strategy is to achieve economies of scale and thus be able to better compete in the global commercial-shipbuilding market. Concentration could create the scale necessary to enable an investment in new fabrication/assembly technologies that would not be possible to justify at a lower scale (lower level of throughput).[5]

Another reason for multisite construction is to most effectively utilise a shipyard's existing assets as the global shipbuilding markets fluctuate. In shipbuilding, as in the other capital-intensive heavy manufacturing industries, efficient capacity utilisation is a key driver of business effectiveness. To this end, partial outsourcing is an effective tool in Japan's shipbuilding environment. What makes it feasible is the somewhat distributed nature of the Japanese shipbuilding industry. Seven major builders all have multiple facilities, and there are more than a dozen medium-sized or second-tier builders (which build large ships just as do the majors). Many medium-sized companies have multiple yards as well. In Japan, labour mobility is low, so subcontractors are used to a much greater extent than in the United States and United Kingdom, to avoid

---

[3]In Japan, this approach has been named Tactical Level Flexible Outsourcing.

[4]In Japan, this is called the Strategic Level Concentration of Investment and Capability.

[5]For example, the new laser steel cutting facility at Bender Shipbuilding (Mobile, Alabama) has productive capacity in excess of the shipyard's needs. Bender was able to make this investment, which improves its shipbuilding capability, by setting up the new facility to do a substantial amount of its work for outside clients.

having to lay people off during periodic slow periods. Some of this subcontracting involves blocks being built by other companies, including fabricators that are not shipyards.

The combination of a multiplicity of yards and other fabricators, and a subcontracting environment, provides a useful degree of operational flexibility at a tactical level that is to handle market fluctuations. When a yard is very busy, it can sub-out more blocks and thereby push more ships through its final assembly stage. When business is slower, a yard can retain more of the added value of each ship contract in-house. This balancing act tends to work because the multiplicity of shipbuilding companies, shipyards, and steel fabricators confers a degree of flexibility at the total-industry level. It is one of the benefits of Japan's two-tier structure in shipbuilding (Koenig, Narita, and Baba, 2001).

POTENTIAL DISADVANTAGES OF MULTISITE CONSTRUCTION

Multisite construction also involves possible disadvantages or additional costs.

First, the problems of accuracy control become more acute because design and build tolerances must be maintained at several shipyards. Common nomenclature, techniques, and software packages must be used to ensure that the blocks built at different shipyards align correctly during assembly. Problems with alignment can lead to potential significant rework costs.

Blocks must be constructed or reinforced in a way to ensure that dimensional tolerances are maintained during transportation. They also may require additional bracing or structures for the transportation process, which will incur additional costs. Additional costs will also be incurred by maintaining separate trades and workforce for transporting blocks between sites.

Finally, since processes must be coordinated among several shipyards, management of the schedule for construction and delivery of the blocks becomes more difficult. Delays in block construction at one shipyard, or delays in delivery caused by transportation problems, can seriously throw off the schedule for the delivery of the ship.

CONCERNS FOR THE TYPE 45 PROGRAMME

To summarise what we have learned from the survey of past and current ship construction processes, building the Type 45 ships from large blocks at multiple shipyards is technically feasible. As described in the body of this book, overall programme costs with the block strategy should be lower than having each shipbuilder construct entire ships and only approximately 4 percent greater than having sole-source production for the Type 45 class.[6] Both BAE SYSTEMS Marine and VT will maintain their ability to build warships and, therefore, should provide competition for future MOD programmes. Finally, the block strategy will facilitate the VT move to Portsmouth, thereby reducing overhead costs at the Portsmouth repair facility and providing new facilities for the construction of ships in the United Kingdom.

However, building, transporting, and assembling blocks of the size and complexity envisioned for the Type 45 programme have never been undertaken by either BAE SYSTEMS Marine or VT. Furthermore, the two companies have not previously worked together on the construction of a ship. Management control and coordination between the two organisations are extremely important to ensure that structural tolerances of the blocks are maintained at the two shipyards, both during production and after transporting the blocks. Also, most previous examples of multisite construction have been for commercial shipbuilding programmes. Naval ships, especially warships, are more complex and require more system and communication interfaces. These requirements will magnify the problems with aligning blocks.

Having both BAE SYSTEMS Marine and VT participate in the design of the Type 45 is an important determinant of success. This joint design effort must ensure that the necessary engineering and build information is prepared in the correct sequence and in a manner that is understood by all the shipyards involved in the programme. The information must also specify the materials and equipment needed

---

[6]Note that our cost estimates for the block options did not include significant rework if blocks did not align properly. Therefore, our cost estimates for the block options are somewhat conservative.

by the various shipyards for the construction of their blocks and describe the source and schedule for the material and equipment.

During construction of the blocks at each shipyard, quality-control functions must examine the blocks during construction and identify any potential distortion problems when the blocks are completed. Detailed finite-element analysis is also required to understand the requirements for and potential effects of proposed lifting and transportation plans. All software tools, procedures, nomenclature, and methods must be coordinated among all the shipyards to avoid problems in matching the blocks during final assembly.

All appropriate costs for building and transporting the blocks must be identified and factored into the total Type 45 programme costs. For example, the weight of the blocks being transported is much larger than has typically been moved over water. Special barges/ships will be required, which could add substantial costs. The outfitted blocks will have to be made weather-tight, not only for the sea voyage but also in the shipyards while they are being outfitted initially and prepared for joining together. An aspect that must be considered is the change in block size due to temperature. BAE SYSTEMS Marine will be required to measure the joining edges of the block received at Barrow to eliminate any inaccuracy due to temperature difference between Portsmouth and Barrow. With the planned block division for the Type 45, such inaccuracies could be up to an inch in breadth and depth. Finally, the shipyards receiving and assembling the blocks must have the capability to receive, transport, and load the blocks onto the assembly berth. Creating this capability at shipyards that do not have it could add significant costs to the programme, thereby negating any total cost benefit.